Interpretation
of
Geometric
Dimensioning
and
Tolerancing

SECOND EDITION

Interpretation
of
Geometric
Dimensioning
and
Tolerancing

SECOND EDITION

Daniel E. Puncochar

Industrial Press Inc.
200 Madison Avenue
New York, New York 10016-4078

Library of Congress Cataloging-in-Publication Data

Puncochar, Daniel E.
    Interpretation of geometric dimensioning and tolerancing / Daniel
E. Puncochar. — 2nd ed.
    128 p.
    Includes index.
    ISBN 0-8311-3010-5
    1. Engineering drawings—Dimensioning.   2. Tolerance (Engineering)
I. Title.
T357.P96   1996
620'.0045—dc21

96-37138
CIP

Industrial Press Inc., 200 Madison Avenue, New York, New York 10016-4078

Interpretation of Geometric Dimensioning and Tolerancing, Second Edition

9 8 7

TABLE OF CONTENTS

PREFACE

Geometric dimensioning and tolerancing (GD&T) is a method for stating and interpreting design requirements. GD&T is an international system of symbolic language, and is simply another tool available to make engineering drawings a better means of communication from design through manufacturing and inspection. Some of GD&T's advantages are uniformity in design practice, fewer misinterpretations, ensured interchangeability, and maximum tolerance allocation. Also, with GD&T, design requirements are specified explicitly and the latest gaging techniques are accommodated. These advantages contribute to higher production yields with less rework or scrap.

To help the reader understand GD&T, *Interpretation of Geometric Dimensioning and Tolerancing* begins with basic principles and builds on these principles with applications-oriented concepts. Complex material is presented in a "building-block" approach, with examples that illustrate each concept. End-of-chapter evaluations further reinforce the explanations given in each section. It is assumed that each reader has an adequate knowledge of production drawing reading. Because GD&T applies regardless of the measurement system used, some drawing examples are dimensioned and toleranced in inches and some in millimeters.

Interpretation of Geometric Dimensioning and Tolerancing completely covers the material in ASME Y14.5M-1994, but does not prescribe design practices, state design requirements, specify inspection techniques, or specify any other engineering practice. However, it is sometimes necessary to state how something is specified or inspected so that a concept can be discussed adequately. In addition, the drawings in this text are not complete production drawings, but only present the concepts currently under discussion.

It is hoped that *Interpretation of Geometric Dimensioning and Tolerancing* will assist the reader in becoming conversant in the techniques of GD&T given in the ASME standard—techniques that can be integrated smoothly into engineering design systems and modern inspection systems.

ACKNOWLEDGMENT

The second edition of this book is published primarily as the result of two very special people, my wife and son. They both gave me the encouragement needed to complete the revision.

Donna, my bride, patiently revised, edited, reorganized as required and assisted tremendously with the indexing of this edition. Her meticulous methodology is invaluable. Her hours of work did not go unrecognized.

And to my son, Michael, who made this material easier to comprehend through the new and revised illustrations. Without illustrations, this book would only be an organized group of words.

INTRODUCTION

History

During the early period of manufacturing there seldom were any drawings. A person had an idea for something that was needed for industry, farming, or mining, and made it. Usually there was only one item, and when repair was needed, someone repaired or replaced the needed part right on the job.

Over time, complexities in manufacturing increased, and there was an increased need for drawings of parts and their assemblies. With drawings came tolerances—parts were permitted some variation rather than being fitted to only one assembly. This tolerance was specified as a plus/minus tolerance.

This plus/minus or coordinate system worked quite well and still does for many applications. But today there is a need for interchangeability of parts and assemblies manufactured around the world being brought together at an assembly plant that makes automobiles, tractors, airplanes, televisions, projectors, spacecraft, and so on. These items also need replacement parts that assemble readily without a lot of rework.

As the demand for parts manufactured around the world grew, so did the need for accuracy. Accuracy became more critical because of competition for parts and assemblies. The idea of positional tolerancing, which provided a means for locating round features within a round tolerance zone rather than the traditional square tolerance zone, was introduced. The idea caught on and was adopted by the military. It became part of the military standards and later was a Unified American Standard Association standard, ASA Y14.5. This standard was released in 1956 and was accepted by the military. Later, ASA became the American National Standards Institute (ANSI). ANSI later published a complete system of symbology for geometric form and positional tolerances, "Dimensioning and Tolerancing."

In 1983, ANSI Y14.5M-1982 was released. This standard clarifies some of the old practices and moves a little closer to the practices of the International Organization of Standards (ISO). ISO is primarily a European standard. In 1995, ASME Y14.5-1994, from the American Society of Mechanical Engineers, was released. This new standard further clarifies requirements within the standard and also moves more in line with the ISO standard. Today, geometric dimensioning and tolerancing (GD&T) is used by the majority of manufacturing companies in the United States and the world.

The Importance of Standards

We live and work in accordance with standards all of the time. Almost every aspect of life, education, or business operates according to some standard. Some of the standards are specified and controlled locally, whereas others are national and still others are international.

WHAT IS A STANDARD?

A standard is a model or rule with which other similar things are to be made to or compared; GD&T symbols, principles, and rules are the model that is provided internationally. This system was created to improve communications, control, and productivity in manufacturing throughout the world. Standards are critical to all of us, and they will become increasingly important as our technologies continue to develop.

CHANGE

Change is one of the most constant things we have as a society. As a result, standards change also, especially those associated with technology. GD&T is an example of a widely used standard that must be updated constantly to be useful to industry. Over time, needed improvements are identified, discussed, and evaluated. When a sufficient amount of improvements are agreed upon, standards are changed. The most recent update to the GD&T standard was adopted in 1994.

UNIVERSALITY

GD&T is a standard throughout the world. In the United States, the standard is maintained by the American Society for Mechanical Engineers (ASME). In the European countries and other parts of the world, the standard is maintained by the International Organization for Standards. These two standards are not identical, but with each revision the standards become more similar. As our international companies continue to exchange drawings among themselves, these standards will become increasingly important.

Why GD&T

GD&T adds clarity and contributes its many advantages to our coordinate system of dimensioning. The old system (coordinate) was lacking in many respects. A part of the designer's intent was always left to interpretation by the craftsman (i.e., dimension origin, form profile and orientation). Probably the most significant difference between the two systems is the location of round features; the coordinate system had a square tolerance zone, which allowed some good parts to be rejected. In our world of high technology, high cost, and transfer of parts around the world, we cannot tolerate the misinterpretation that is possible with the coordinate system.

GD&T IS NOT A REPLACEMENT

The coordinate system is not being replaced entirely with GD&T. GD&T is specified to enhance the coordinate system as required per design. When the advantages of GD&T can be utilized, they are simultaneously specified. GD&T, a system of symbols, provides a means of completely specifying uniformity and describing the designer's intent. These symbols eliminate most of the drawing misinterpretation by not having notes in drawing margins and by having complete descriptions of features and design requirements.

COMPLETE SPECIFICATION

A complete specification of design requirements is possible with symbols. These symbols also allow the designer to specify maximum tolerances for parts that must assemble with other parts. These maximum tolerances also ensure the interchangeability of parts. The use of symbols for complete specification is becoming increasingly important with the growing interrelated ownerships of companies around the world. GD&T is fast becoming an international "common symbolic language" controlled by standards. Today, the majority of U.S. manufacturing companies are applying GD&T to their drawings.

ADVANTAGES

There are many reasons for specifying geometric tolerancing wherever design integrity must be controlled and communicated completely to others. Two key principles for applying GD&T are the function and the relationship of parts in an assembly. Probably the most advantageous part of GD&T is the method of specifying feature location. In the past, features were located with the coordinate system.

The coordinate system is a method of tolerancing that uses a plus/minus tolerance. Plus and minus tolerances are specified for lengths, widths, diameters, shapes, and locations. An illustration of how drawings may be dimensioned and toleranced with the coordinate system is shown in Figure 1-1. This method of tolerancing permits the length and diameters to vary

Figure 1-1

Figure 1-2

Figure 1-4

Figure 1-5

by a plus/minus value. It also allows the maker of the part to put the center hole wherever he or she desires. By looking at the drawing, we can only assume that the hole is centered. This is an example of a drawing being left open to misinterpretation by anyone reading it.

A similar situation exists where hole or pin locations are specified with the coordinate system. An example of how holes are specified is shown in Figure 1-2.

The tolerance as specified establishes a square tolerance zone based on the plus/minus five thousandths of an inch tolerance in the *x* and *y* directions. There is no consideration for the actual mating size. The tolerance zone is ten thousandths of an inch on each side of a square regardless of the actual mating size. An illustration of how the tolerance zone appears is shown in Figure 1-3.

The axis of the hole or pin must be positioned in that square zone in order for the feature to be located properly. The feature may lean or slant an uncontrolled amount as long as the axis stays within the square zone. The designer only assumes that the feature will be

produced nearly perpendicular to the material it is put into. If the axis were in an extreme corner of this zone, the feature location is still acceptable; that radial measurement is .007 in., as shown in Figure 1-4.

The only place there is a .007 in. measurement is from the center to any corner. That .007 in. should be usable all around the desired true position, as illustrated in the example in Figure 1-5. GD&T provides a method of specifying a tolerance zone that takes the shape of the feature into consideration if so desired by the designer. GD&T also allows consideration for the feature's actual local size for calculating total tolerance. This concept is presented in later chapters.

Other advantages of GD&T include international uniformity in describing designer intent. The symbolic method of specifying design intent eliminates most misinterpretation of drawing notes. In the past, drawings usually contained in the margin a list of notes that were intended to explain certain requirements. These notes were all subject to misinterpretation. With the available symbols, designers can more readily specify complete design requirements. The proper application of geometric tolerances ensures the interchangeability of parts. GD&T is very rapidly becoming a common language throughout industry internationally.

Figure 1-3

Chapter 1 Evaluation

1. Drawings are the primary _____ tool between designers and manufacturing.

2. GD&T is a system made up of _____ primarily.

3. The GD&T standard is one of the standards maintained by the _____ is required.

4. GD&T adds to the coordinate dimensioning system when specific _____ is required.

5. GD&T does not _____ the coordinate dimensioning and tolerancing system.

6. GD&T can best be described as a _____ dimensioning and tolerancing system.

7. GD&T is used to control the _____ of a part feature and its relationship to other features.

8. GD&T is also used to control feature _____ and _____.

9. The key principles of GD&T are _____ and _____.

10. Two advantages of the GD&T system are maximum _____ and ensured _____ of mating parts.

11. The total amount that a part size may vary is a size _____.

12. A common method used to specify a tolerance for the nominal size of a feature is _____ values.

SYMBOLS AND ABBREVIATIONS

Introduction

Geometric dimensioning and tolerancing (GD&T) is a language of symbols. This chapter will introduce these symbols and give an application for each of them. Throughout the remainder of this book you will see these symbols applied to various drawings in combination with other symbols. The symbols presented here are the ones used to specify geometric characteristics and dimensional requirements on industrial drawings, which are in accordance with ASME Y14.5M-1994.

One of the advantages of GD&T is that it is an international language of symbols that generally eliminates the need for drawing notes. (See Appendix B for the symbols.) However, in limited situations it may be necessary for the designer to add a short note to aid in conveying the design requirements. In Figures 2-1 and 2-2 are two examples of when a note may be specified.

Figure 2-1

Figure 2-2

Symbols

DIAMETER

The term is very commonly used with many aspects of our lives. Most of us are familiar with the abbreviations D and DIA. Now with GD&T, we also have a symbol for diameter; it is a circle with a slash through it, ∅. The diameter symbol is used to describe cylindrical features and tolerance zones.

This symbol always precedes the feature size or tolerance specification. Figure 2-3 is an example of how this symbol appears in application. **Note**: The absence of this symbol before either the feature size or tolerance indicates a noncylindrical feature or tolerance.

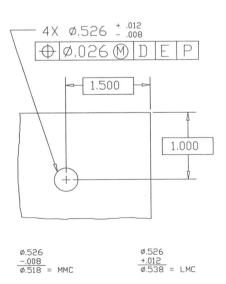

Figure 2-3

BASIC

Basic is a term used to describe the theoretically exact size, shape, or location of a feature or datum. Basic dimensions do not have a tolerance. Tolerances for basic dimensions are specified in feature control frames, or tooling tolerances apply under other conditions. Basic can be abbreviated as BSC, or, most recently, the use of the basic symbol is used; the symbol for basic is a rectangle around the dimension.

Basic is used to describe the theoretically exact shape or profile of surfaces either regular or irregular. Most frequently, basic is used to specify the exact desired position of features. Then, each feature is given a tolerance that allows some variation depending on the design requirement. Figure 2-3 shows an example of how basic may be specified.

MODIFIERS

The next symbols and abbreviations to be introduced are those for the modifiers. The modifiers are maximum material condition (MMC) and least material condition (LMC). These may modify a specified tolerance, as their name indicates. Regardless of feature size (RFS) is implied for all geometric tolerances. The symbol does not exist in the ISO Standard. There is an alternative practice in the ASME Y14.5M-1994 Standard that permits the specification of RFS for a tolerance of position. The first modifier to be discussed in this chapter is maximum material condition.

The symbol for maximum material condition is a capital "M" in a circle, Ⓜ. Maximum material condition may also be abbreviated MMC. Most frequently, the symbol is specified on the drawing, and the abbreviation is used during discussions about tolerancing.

The term maximum material condition is used to describe the maximum condition of the actual mating size. For example, a hole is a feature of size that is permitted to vary in size within the limits of a plus/minus size tolerance. For holes or any internal feature, MMC is the smallest actual mating size for that feature. In other words, the maximum material remains for the piece the hole was put in. An example of a hole size specification is shown in Figure 2-3. When the hole is produced at its smallest actual mating size Ø.518, the material is at MMC.

Figure 2-4

Maximum material condition is also applicable to external features of size, such as pins, tabs, and splines. With these features MMC is equal to the largest actual mating size permitted by the size specification. An example of a pin size that permits an MMC of Ø.512 is shown in Figure 2-4. Maximum material condition can be thought of in terms of the smallest hole and largest pin, or it can be thought of in terms of weight; smallest holes—more weight, largest pins—more weight in part.

The application of this modifier would include designs where clearance is required for assembly. Maximum tolerances are achieved as the feature departs from MMC. A minimum tolerance is stated and then a modifier symbol is added following the tolerance when applicable to permit an increase in tolerance equal to the departure from MMC. This concept will be covered thoroughly in Chapters 6 and 8.

LEAST MATERIAL CONDITION

The next modifier is least material condition. The symbol is a capital "L" in a circle, Ⓛ. As with maximum material condition, the symbol is used on drawings and an abbreviation LMC is used during tolerance discussions.

The term least material condition is used to describe the opposite condition as maximum material condition. This modifier is specified for features of actual mating size to describe the least material condition. For example, by using the same hole callout as in Figure 2-3, the LMC is equal to Ø.538 (Figure 2-5). Here the least amount of material remains in the material after the hole is put in it.

Figure 2-5

Least material condition also applies to external features of size. For external features subject to size variation, LMC means these features contain the least material within the specified size limits. In Figure 2-6, an example of a pin at least material condition is shown.

The application of this modifier would include designs where a part is located, or a minimum material thickness must be maintained, such as holes near material edges or bosses with holes bored in them where the wall thickness is critical. This modifier provides the same advantages as MMC, but in opposite direction.

REGARDLESS OF FEATURE SIZE

The final modifier to be covered is regardless of feature size. It is considered a modifier even though it does not permit any modification of stated drawing tolerances. As the name indicates, regardless of the

features size, the stated tolerance applies.

This modifier is a very restrictive requirement. It applies with respect to the individual tolerance, datum reference, or both where no modifying symbol is specified.

This modifier is applied with tolerances for features of size like the other two modifiers; this one just limits the tolerance to that specified by the designer for parts that align.

Regardless of feature size also applies to external features of size. External features here might be splines or gears and similar types of features that are subject to size variation. With these types of features, designs usually cannot tolerate play or additional tolerance between mating parts.

The application of this modifier would include designs where tolerance allowances are critical. This modifier may be implied for parts that form an assembly with limited shift between parts. Typical applications for restricted tolerances are gears, splines, and press fits.

FULL INDICATOR MOVEMENT

The term full indicator movement is a new term for older terminology. This new term replaces total indicator movement (TIM) and total indicator reading (TIR). There is no symbol for full indicator movement; it is abbreviated FIM. This abbreviation does not appear on drawings. FIM is understood or implied for certain form controls, which control cylindrical features. FIM will be discussed further in Chapter 6. An example of how FIM is used on a drawing is shown in Figure 2-7. Figure 2-7 means that the stated tolerances of .002 and .005 are FIM. For example, FIM is the

Figure 2-6

Figure 2-7

Figure 2-9

Figure 2-8

complete movement of a needle on a dial indicator. To measure the variation of the Ø.500 and Ø1.000 inch diameters, a dial indicator would be rested on each surface and then that surface rotated 360 degrees. During this complete rotation, the dial indicator reading (FIM) must not exceed the stated tolerance.

PROJECTED TOLERANCE ZONE

Projected tolerance zone is not abbreviated; the symbol for it is a capital "P" in a circle, Ⓟ. The projected tolerance zone describes a tolerance, usually for a fixed fastener application, to prevent interference between mating parts.

The projected tolerance zone specifies a tolerance of location through the projected height. As shown in Figure 2-8, the axis of the hole must be controlled within a specified tolerance above the part the hole is in. The tolerance zone is projected to a height equal to the thickness of the mating part. The purpose of this control is to prevent the hole for the fixed fastener from interfering with the mating part.

DIMENSION ORIGIN

The term dimension origin is not abbreviated, and its symbol is ⬦━▶. This symbol is used to identify the surface or feature where the dimension originates.

Some designs are complex, thus difficult to determine where dimensions are to begin. In these situa-

tions, the designer specifies where the dimension is to originate. Usually the part will be different if made by starting dimensions from a different surface of feature than intended. The example in Figure 2-9 illustrates such a part.

Another example of using the dimension origin symbol is in Figure 2-10. Here it specifies where the angular measurement begins when an angle is specified with linear and angular dimensions. This tolerance zone will widen from the apex of the angle as the distance increases.

This symbol may also be utilized to specify a basic angle for a surface. In Figure 2-11, the symbol indicates where the dimension originates to establish the basic 45° angle. This tolerance zone will remain an equal width within parallel boundaries for the length of the feature surface.

ALL-AROUND

All-around is a term that is not abbreviated. The symbol is ↗Ⓞ. This symbol is specified along with profile specifications. It means that the profile tolerance

Figure 2-10

Figure 2-11

Figure 2-13

applies all around the controlled feature and replaces the words "all-around."

This symbol should only be applied when a uniform profile tolerance is required all around the feature; also, the tolerance is the same around the feature. Figure 2-12 shows an example of how the uniform tolerance applies.

RADIUS

Radius, like diameter, is another older term that was used on industrial drawings. Radius is not abbreviated; the letter "R" is used as the symbol.

Radius is applied to designs that require the removal of edges or to rounded features. The letter "R" precedes the radial dimension. When a radius is dimensioned in a view that does not show the true shape of the radius, "TRUE R" precedes the radial dimension. See the application of radius in Figure 2-12. The actual radius must be within the limits of size and must have perfect form with no flats or reversals.

REFERENCE DIMENSION

Another term that is not new to drawings is reference dimension. This term was usually abbreviated to REF and is now represented with a value in parentheses, ( ).

Reference dimensions are specified for a relationship between features in a flat pattern application. It is not used to define parts. Reference dimensions may be the sum of several dimensions, the size or thickness of material, and specify travel of moving parts, etc. See Figure 2-13.

SPHERICAL DIAMETER

This term is abbreviated as SD, the symbol is SØ.

Spherical diameter is specified for round features. It specifies the diameter of these features. The abbreviation or symbol is specified either before or following the round feature size (Figure 2-14.)

SPHERICAL RADIUS

This term may be abbreviated SR; there is no symbol. Spherical radius is applied to round features. The

Figure 2-12

Figure 2-14

SR 6.000
± .010

Figure 2-15

1.000   .625
± .010   ± .010

Figure 2-17

3.250
± .010

Figure 2-16

abbreviation is specified before the radial value of the feature. See the application in Figure 2-15.

ARC LENGTH

Arc length is a term used to describe the length of a curved surface. The symbol is $\overset{\frown}{105}$, and there is no abbreviation. This symbol is placed over the dimension.

The arc length symbol is specified when it is required to measure along the actual part surface. When this symbol is specified, a linear measurement across the arc is not permitted. In the past, terms like "TRUE" or "TRUE ON SURFACE" were used. See the example in Figure 2-16.

CHAIN LINE

Chain line is a term used to describe or identify a specific area, surface, or portion of a part for special treatment. The term has a symbol, which is — - —, and there is no abbreviation.

The chain line symbol is applied when the designer requires only a limited portion of surfaces or areas to be treated differently from the rest of the part. See Figure 2-17.

CONICAL TAPER

The term conical taper was introduced with the 1982 update to the standard. There is no abbreviation; the symbol is ▷.

There are three methods of specifying a conical taper. First, it may be specified with basic dimensions for the diameters and the taper (Figure 2-18). There may be a feature size and profile tolerance combined with a profile of the surface (Figure 2-19). The diameters along with the length may be toleranced (Figure 2-20).

SLOPE

There is no abbreviation, and the symbol is ◁.

Slope is primarily specified to control flat tapers. Slope is not specified as degrees, but as a ratio of height differences from one end of the flat taper to the other end. An example of how slope is specified is shown in Figure 2-21.

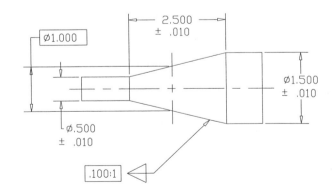

2.500
± .010

Ø1.000

Ø1.500
± .010

ø.500
± .010

.100:1

Figure 2-18

Figure 2-19

Figure 2-22

Figure 2-20

COUNTERSINK

Countersink, like the previous terms, is not new to industrial drawings. Countersink may either be abbreviated as CSink or specified with the symbol ⋁.

Countersinks are specified for features that require a smooth flat surface. Countersinks are typically specified as shown in Figure 2-26.

COUNTERBORE/SPOTFACE

The terms counterbore and spotface arc not new to industrial drawings. The symbol is new, however; it is ⌴. The terms may also be abbreviated as in the past—CBore and SF. The symbol is preferred with a GD&T application.

Counterbore or spotface may be specified for features that require a recessed or flat mounting surface for fasteners. A typical callout for a counterbore or spotface using the symbol is shown in Figure 2-22.

Figure 2-23

Figure 2-21

Figure 2-24

Figure 2-26

DEPTH/DEEP

The terms depth or deep are not new to industrial drawings. These are words that were written out on drawings. Now there is a symbol for them, $\overline{\underline{\vee}}$, or the abbreviation DP is used.

Depth or deep is specified for features that do not pass completely through a part. The symbol may be specified as in Figure 2-22 for counterbores, or it may be specified for blind holes as shown in Figure 2-24.

DIMENSION NOT TO SCALE

At times, it is desirable or necessary to specify dimensions not to scale. The current practice is to use a heavy straight line under the dimension.

The symbol for dimensions not to scale, **15**, is specified when dimensions are intentionally drawn out of scale. Today, with computer graphics, there is seldom a situation where out-of-scale dimensioning is used, however, agreement shall be maintained between the

defining dimension and the graphics presentation of the feature, in all views. There shall always be complete agreement in the feature defining dimension and the true size, location, and direction. An example of out-of-scale dimensioning is illustrated in Figure 2-25.

NUMBER OF PLACES

This term is not really new to drawings; they just have a little different application now. There is no abbreviation; the symbol is an **8X** .

The designer may specify the number of places or times for holes in a pattern, for example, as shown in Figure 2-26. The same symbol is specified when a slot size is specified. See the example in Figure 2-27.

Figure 2-25

Figure 2-27

Figure 2-28

STATISTICAL TOLERANCING

This symbol, ⟨ST⟩, is introduced to indicate that a tolerance is based on statistical tolerancing. There is no abbreviation. When a tolerance is a statistical geometric tolerance, the symbol is placed in the feature control frame following the stated tolerance and any applicable modifier (see Figure 2-28).

Statistical tolerancing may be applied to increase individual feature tolerance. Tolerances for individual features of an assembly are determined mathematically by dividing the tolerances among the features. When this assignment is mathematically restrictive when adding them, a statistical tolerance would be specified. Statistical tolerances on dimensions are designated as shown in Figure 2-29.

A drawing note such as the following must be added to the drawing: **FEATURES IDENTIFIED AS STATISTICALLY TOLERANCED ⟨ST⟩ SHALL BE PRODUCED WITH STATISTICAL PROCESS CONTROLS (SPC).**

If it is necessary to specify both the statistical limits and the arithmetic stacking limits (Figure 2-30) when the feature may be produced without SPC, a note such as the following must be added to the drawing: **FEATURES IDENTIFIED AS STATISTICALLY TOLERANCED ⟨ST⟩ SHALL BE PRODUCED WITH STATISTICAL PROCESS CON-**

NOTE:
FEATURES IDENTIFIED AS STATISTICALLY TOLERANCED ⟨ST⟩ SHALL BE PRODUCED WITH STATISTICAL PROCESS CONTROLS, OR TO THE MORE RESTRICTIVE ARITHMETIC LIMITS.

Figure 2-30

TROLS, OR TO THE MORE RESTRICTIVE ARITHMETIC LIMITS.

Statistical tolerancing may also be specified for other features such as holes. Figure 2-31 illustrates how statistical tolerancing may be applied to a hole specification.

CONTROLLED RADIUS TOLERANCE

The abbreviation for a controlled radius is CR. There is no symbol. Radius is specified when a fair curve without reversals is desired. The zone is created by two arcs tangent to the adjacent surface (see Figure 2-32). The actual feature surface must lie within the tolerance zone. The difference between a radius and a controlled

NOTE:
FEATURES IDENTIFIED AS STATISTICALLY TOLERANCED ⟨ST⟩ SHALL BE PRODUCED WITH STATISTICAL PROCESS CONTROLS.

Figure 2-29

Figure 2-31

Figure 2-32

radius is that a radius may have reversals; the only requirement is that the produced feature must fall within the tolerance zone.

BETWEEN

There are designs where the tolerance for the feature applies to only a portion of the feature. In these instances, the designer has the between symbol available, ⟷. There is no abbreviation. The arrowheads may or may not be filled. An example of the between symbol is shown in Figure 2-33.

FREE STATE

When an individual form or location tolerance is applied to a feature in the free state, the symbol, a capital "F" in a circle, Ⓕ, is placed in the feature control frame following the specified tolerance and any modifier, Figure 2-34. There is no abbreviation. This symbol may be specified for parts such as plastic, nylon, rubber, etc., that are not rigid enough to hold their form.

TANGENT PLANE

When a specified tolerance applies to the tangent plane of a feature, the symbol, a capital "T" in a circle, Ⓣ, is placed in the feature control frame following the specified tolerance for the feature, Figure 2-35.

Figure 2-33

Figure 2-34

Figure 2-35

There is no abbreviation. The plane contacts the high points of the surface and shall lie within two parallel planes that are the stated toleranced apart.

SQUARE

Square features may be identified with a □ symbol. There is no abbreviation. This symbol may be specified on a drawing like Figure 2-36 to indicate the feature is square.

Summary

This completes your introduction to the majority of the symbols and abbreviations associated with geomet-

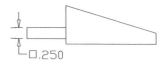

Figure 2-36

ric dimensioning and tolerancing. There are other symbols that will be introduced in other chapters. Those chapters cover datums, form/orientation controls, and tolerances of location.

The symbols and abbreviations introduced in this chapter will be applied to other drawings and examples throughout this book. They were introduced in this chapter as a basis for the other material.

Chapter 2 Evaluation

Fill in the abbreviations and symbols that match the phrases. The abbreviations and symbols may be used more than once and some of them not at all.

_____ 1. The abbreviation for full indicator movement.

_____ 2. The condition of an internal feature when it measures the largest size within design limits or weighs the least.

_____ 3. The symbol specified to indicate diametrical or circular features.

_____ 4. Tangent plane.

_____ 5. Dimensions that are for reference only.

_____ 6. Dimensions that do not have tolerances.

_____ 7. The symbol to indicate diametrical tolerance zones.

_____ 8. The symbol specified to indicate the number of times or places, or by.

_____ 9. A curved surface that is to be measured along the curve is specified with which symbol?

_____ 10. Dimensions that are not to scale are identified with which symbol?

_____ 11. Parts or features requiring a rounded edge or corner are identified with which symbol?

_____ 12. Features that are spherical shaped are identified with which symbol?

_____ 13. The symbol used to indicate the depth of a feature.

_____ 14. The condition of an external feature when it measures the largest or weighs the most.

_____ 15. A slope or flat taper.

_____ 16. Symbol for conical tapers.

_____ 17. To indicate a countersink.

_____ 18. Indicates the origin of a dimension.

_____ 19. To control the perpendicularity of a fastener to a given height.

_____ 20. A counterbore or spotface is specified with which symbol?

_____ 21. Dimensions that specify theoretically exact size, shape, or location of features.

22. The total movement of a dial indicator needle.

1. X	16. Ⓟ
2. LMC	17. FIR
3. SR	18. ⌀(A1)
4. SD	19. (50)
5. ⌀	20. —
6. 2.125	21. MMC
7. R	22. ⌒105
8. ↧	23. ∨
9. ⌒	24. ⌒
10. NO SYMBOL OR ABBR.	25. ⌀
11. ▷	
12. ⌀	
13. RFS	
14. ▷	
15. ∥	

DATUMS

Introduction

Datums provide the framework from which part features are manufactured, and they are critical for all industrial drawings that are created for today's production. Parts, assemblies, and processes are becoming more critical for profitability. The methods and practices that were acceptable in the past are no longer acceptable with today's technological advancements. Parts produced today must be 100% interchangeable with their counterparts. The designer can ensure complete fit, function, and interchangeability by clearly specifying the design intent. Datums provide the clarity required for proper feature orientation, which originates from datum planes. Datums are specified on drawings so that repetitive measurements can be made from design through the production and inspection processes. Datums are nothing more, nor less, than physical features of parts to make repeatable measurements from. These features are considered theoretically exact.

Datum Identification

Datum identification is required with ASME 14.5M-1994. Previously, implied datums were permitted. Too often with implied datums, the same surfaces, edges, or features were not used throughout the design, manufacturing, and inspection processes. This can be illustrated with a simple part. A drawing specification for a simple flat part with three holes in it is shown in Figure 3-1.

This part may not be oriented the same through manufacturing and inspection, causing it to be rejected. To illustrate, consider that the part was manufactured using the left-hand end as the surface to butt up with a fence, and the lower edge against a stop. The holes

Figure 3-1

were put in the part while it was set up as illustrated in Figure 3-2.

Then the part was set up for inspection using the longest surface to make contact first. The part would be inspected as shown in Figure 3-3. When inspected by being set up this way, the part was rejected because the lower-left-hand hole was out of design specification.

Therefore, datums must be specified so the drawing is interpreted the same by all who read it.

Figure 3-2

Figure 3-3

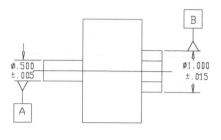

Figure 3-4

WHAT IS A DATUM?

A datum is a theoretically exact line, surface, point, area, or axis that is used as an origin for dimensions. These regions are considered perfect for orientation purposes only. During machining processes, the part is resting against a theoretically perfect or exact datum surface where features of the part have been identified as datum features; the part is oriented and immobilized relative to the datum reference frame in their selected order of precedence. This orientation makes the geometric relationships that exist between features measurable. True geometric counterparts of features used to establish datums may be:

a) a plane
b) a maximum material condition boundary (MMC concept)
c) a least material condition boundary (LMC concept)
d) a virtual condition boundary
e) an actual mating envelope
f) a mathematically defined contour

It is from these actual surfaces that measurements are made to check feature relationships. On the drawing, these datum features are identified with the datum feature symbol.

DATUM FEATURE SYMBOL

The datum feature symbol is a square box that contains a capital block letter with a leader connecting it to the feature with a triangle. The triangle may be filled or not filled. Figure 3-4 shows a drawing illustrating the proper symbol and attachment. Any of the letters of the alphabet may be used except for I, O, and Q, which may be confused with numbers. The letters may be used in any order because alphabetical order is meaningless in this system. The important mental distinction that must be made is that a datum is theoretically perfect whereas the datum feature itself is imperfect.

Three-Plane Concept—Flat

Theoretical datum planes or surfaces are established from a perfect three-plane reference frame. This frame is assumed to be perfect with each plane oriented exactly 90° to each other, referred to as the datum reference frame. This reference frame, with mutually perpendicular planes, provides the origin and orientation for all measurements. When physical contact is made between each datum feature and its counterpart in associated manufacturing or inspection equipment, measurements do not take into account any variations in the datum features. These planes are identified as the primary, secondary, and tertiary datum planes.

PRIMARY

The primary datum is the one that is, functionally, usually the most critical feature or surface on the part. It is part to part interface, typically the largest area when area is involved. The primary datum feature must make contact with the theoretically exact datum plane in a minimum of three points not in a line. The required contact is to prevent the part from "rocking" during manufacturing or inspection processes.

Figure 3-5

This three-point contact is not difficult to achieve; if the designer has any concern about excessive surface irregularity, a surface control may be specified (see Chapter 6). Figure 3-5 shows an example of the primary datum plane establishment.

SECONDARY

The secondary datum plane must be at a 90° angle to the primary datum plane. The secondary datum feature is usually selected as the second most functionally important feature. This feature must be perpendicular to the primary datum feature. There is only a two-point minimum contact required for this plane. These two points establish the part in the other direction to prevent it from rocking about the primary datum plane. This plane may be a stop, fence, or angle plate on processing or inspection equipment. The illustration in Figure 3-6 shows the secondary datum plane.

TERTIARY

The tertiary datum plane must be at exactly a 90° angle to the primary and secondary datum planes. The tertiary or third datum plane is also perpendicular to

Figure 3-7

the other two planes. The part must contact this plane at least at one point. This contact is required for dimension origin and to prevent any back-and-forth movement in the third plane. The tertiary plane could be a locating or stop pin in a processing or inspection process.

All measurements, setups, and inspections are to be made from these three mutually perpendicular planes. Figure 3-7 is an illustration of the theoretical datum reference frame.

The fixture that could be manufactured to orient this part might look like the one illustrated in Figure 3-8. The part must contact three points for a primary datum. Often flat parts similar to this may rock if placed on a machine bed or inspection table; the three raised areas will prevent the rocking. The raised areas are not used to check the hole locations. The secondary datum plane requires a two-point contact, which in this illustration are the sides of the pins. The tertiary datum plane must be in contact with the part in one place only.

Figure 3-6

Figure 3-8

Datum Targets

Datum targets are also used to establish the datum reference frame. Datum targets are used to establish orientation on irregular parts. They may also be used for large surfaces where it would be impractical to use an entire surface as the datum because of size or it is subject to warping, bowing, or other distortions. Usually datum targets are specified on castings, forgings, or weldments, or on any other application where it may be difficult to establish a datum. Datum targets may be points, lines, or areas of a part that provide an orientation-dimension origin. A datum target symbol is used to identify the datum planes.

DATUM TARGET SYMBOL

The datum target symbol is a circle divided in half by a horizontal line. When the symbol is used to identify a circular target contact area, the top half will contain the diameter. Where it is impractical to specify the circular target area inside the upper half of the datum target symbol, the area may be specified outside the symbol and attached to the upper half with a leader line (see Figure 3-9).

The contact area diameter is basic. When the symbol is used to identify any other datum target, the upper half remains open (see Figure 3-10). The lower half of the symbol always contains the datum target identification. The identification consists of a letter (the datum reference letter) and target number. Sufficient targets are specified to satisfy the three-plane datum concept and required points of contact.

Datum target symbols are attached to the datum target with a leader line. A solid leader line indicates the datum target is on the near side of the part; a dashed or hidden line type leader line indicates that the datum target is on the back or far side of the part (see Figure 3-11).

Figure 3-9

Figure 3-10

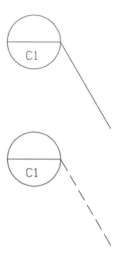

Figure 3-11

Datum targets are dimensioned with basic dimensions or toleranced dimensions (see Figure 3-12). When basic dimensions are specified, tooling tolerances apply. Basic dimensions, as you learned earlier, do not have tolerance, but to aid in locating datum targets properly some tolerance is required. Therefore, tooling tolerance (usually one-tenth of specified feature tolerance) is permitted. The feature tolerance used to calculate the tooling tolerance is the specified tolerance for the feature oriented from the datums. Usually basic dimensions are specified because of subsequent machining and inspection processes.

DATUM TARGET—POINT

A datum target point is identified with an "X," as shown in Figure 3-12. The "X," or datum target point, is usually located dimensionally with basic dimensions. The target point regularly identified is a front view of the part. Where there is no front view, the point location is dimensioned on two adjacent views. Datum target points are normally simulated or picked up with the point of a cone or spherical radius pin.

Figure 3-12

Figure 3-14

DATUM TARGET—LINE

The datum target line may be identified with either a phantom line or an "X." The phantom line is shown in the front view and "X" in an adjacent view. These symbols are located with basic dimensions, as shown in Figure 3-13.

Datum target lines are simulated with the edge contact of pins.

DATUM TARGET—AREA

The datum target area is identified with a phantom line circle, with section lines inside the circle. The area is specified in the upper half of the datum target symbol. If it is a circular area, the size is preceded with the diameter symbol. Phantom lines and dimensions are specified to define the size and shape of target areas.

Figure 3-13

Dimensioning of diameters is not required; the datum target symbol specifies the target area diameters (see Figure 3-14).

Datum target areas are simulated with a flat nose pin. In the case of diameters, the pin is the diameter specified in the datum target symbol, using gage tolerances for the pin diameter.

DATUM TARGETS—OUT OF DIRECT VIEW

Occasionally it may be necessary to identify a datum target point or line without a direct drawing view. Some design requirements may force the datum identification to other drawing views. Figure 3-15 shows point target dimensioning.

When the datum target line is specified out of the direct view, the locating dimensions are only specified in one direction. In Figure-3-16, the preceding part with a datum target line is shown.

DATUM TARGETS—COMPLEX OR IRREGULAR SURFACES

When datums are identified on complex or irregular surfaces, the datum feature symbol should be attached only to identifiable datum features. When datums are established by datum targets, the datum feature symbol is not required. In this example (see Figure 3-17), like many complex parts, no surface can be identified as a datum feature. The datum feature symbol may be applied to extension or center lines as appropriate.

Figure 3-15   Figure 3-16

INCLINED DATUM FEATURES

When the datum reference frame includes an angular feature, the angle of that feature must be specified as basic. Then to position the part in relation to the specified datums, a true contacting plane must be oriented at the basic angle of the feature (Figure 3-18).

Three-Plane Concept—Circular

Circular parts, like noncylindrical parts, also require a three-plane concept for repeatable orientation. The primary datum plane is frequently one flat surface of the part. Then two planes (X and Y), intersecting at

Figure 3-17

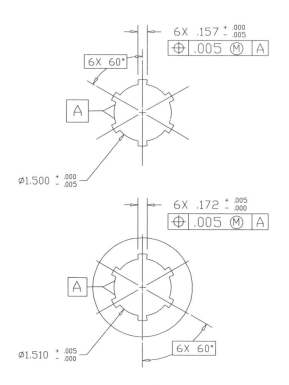

Figure 3-18

Figure 3-20

right angles, establish the axis. This axis is then used as the theoretically exact datum axis. The two intersecting planes provide dimension origins in the *x* and *y* directions for related part features. Refer to Figure 3-19.

There are only two datum features referenced for a part like the one in Figure 3-19. The primary datum plane is one datum feature and the other is the intersection or axis of the *X* and *Y* planes. All dimensioning,

orientation and measurements originate from the axis and planes.

The drawing in Figure 3-20 is an example of how the three-plane concept applies to a circular part. Even though the overall diameter of the part is shown as datum "A," only the theoretical axis is used for orientation of related features. Also, basic dimensions are used to locate features from the theoretical planes.

Partial Datums

Occasionally designs require a datum on a particular surface, but not necessarily the entire surface. Examples of such situations would be large parts, weldments, castings and forgings, and plastics. Some designs incorporating these parts will have partial datums specified. A partial datum is specified to reduce special treatment to an entire surface such as machining or controlling straightness or flatness. A partial datum is specified with a chain line symbol and cross-hatching of the datum area as illustrated in Figure 3-21.

Figure 3-19

Figure 3-21

Figure 3-22

Partial Datums— Mathematically Defined

Sometimes compound curves and contoured surfaces are required to be datums. These features can be used for datums only when they can be defined mathematically and related to a three-plane datum reference frame. When such surfaces are used as datums, the theoretically true geometric counterpart of the shape is used to establish the datum.

Datums of Size

A datum of size is any feature subject to actual mating size variation based on size dimensions. A feature of size is a hole, slot, tab, pin, etc. Because variations are allowed by the size tolerance, it becomes necessary to determine whether RFS, MMC, or LMC applies in each case. A datum feature of size is not a single point, line, or plane. Features that are datums and subject to actual mating size variation must be verified with a simulated datum (Chapter 5, Rule 4).

Figure 3-22 is a drawing of an external feature of size. The diameter of this part is subject to actual mating size variation. When features of this type are specified as datums, the material condition must be specified with the datum identification letter in the feature control frame (Chapter 5, Rule 2). The effect of material condition and datum features of actual mating size variation is explained in detail in Chapter 8.

EXTERNAL CYLINDRICAL

External features of size are verified with an adjustable chuck, collet, ring gages, etc. They are used to simulate a true geometric counterpart of the feature and to establish the datum axis or center plane. The geometric counterpart (or actual mating envelope) is the smallest circumscribed cylinder that contacts the datum feature surface that determines the simulated axis. The axis of these irregular features must be established so that measurements and feature relationships may originate from them. The axis of the simulated datum (true geometric counterpart) becomes the datum axis for all related dimensions. This axis can be determined with a height gage, coordinate measuring machine, or any other similar instrument. External features are simulated, as illustrated in Figure 3-23.

INTERNAL CYLINDRICAL

Internal features of size are verified in a similar manner. If the feature of size is a hole, the true geo-

Figure 3-23

Figure 3-25

Figure 3-24

metric counterpart or actual mating envelope is determined by the largest inscribed cylinder that will fit the hole. The cylinder must be an expandable pin, mandrel, gage, etc. These are utilized to simulate a true geometric counterpart of the feature. The geometric counterpart axis becomes the datum axis for dimension origins and location of related features. This true geometric counterpart axis may be determined with a height gage, coordinate measuring machine, or similar instrument.

Figure 3-24 illustrates how the true geometric counterpart axis is determined.

NONCYLINDRICAL

Internal features may also be specified as datum features. These features are subject to actual mating size variation and must be simulated with a true geometric counterpart to determine the datum plane, center line, median plane, etc. The true geometric counterpart plane is determined with two parallel planes separated to make contact with the corresponding surfaces of the specified datum feature. The true geometric counterpart may be a gage block, an adjustable gage, or measuring instrument used to establish the datum plane, center line, etc.; measurements then originate from the true geometric counterpart. Figure 3-25 is an illustration of a datum feature simulated with a gage block.

Datum features of size must also have modifiers specified for them when associated with the positioning of features (Chapter 5, Rule 2). An example of modifiers being specified is illustrated in Figure 3-26.

When a datum reference letter ("B" in this example) is followed with a modifier, additional consideration must be made for that datum feature. According to the datum/virtual condition rule (Chapter 5, Rule 4),

datum feature "B" must be used at its virtual condition even though it is modified to MMC.

Datum reference letters identifying features of size are implied RFS if not modified with MMC or LMC and must be treated like any other datum feature of size. The datum is a true geometric counterpart established by an adjustable gage to contact the datum feature as produced.

Pattern of Features

Feature patterns may also be used as datums. For example, a pattern of holes may be used as a datum to locate other functionally related features. In Figure 3-27, the pattern of holes is at MMC and as a group establishes datum "A." Each hole has a datum axis established at true position for each hole. These axes of true cylinders simulate the virtual condition of the holes. When the part is resting on the primary datum surface, the hole pattern establishes the second and third datum planes of the datum reference frame. The axis of the pattern of holes (X and Y planes) may depart from the axis of the datum reference frame as the datum feature departs from MMC.

Summary

Datums are assumed to be theoretically exact in order to ensure repeatability from design to inspection. Datums are dimension origins used to establish mea-

Figure 3-26

Figure 3-27

surements and feature-to-feature relationships. Datum features, on the other hand, are actual part features that include all variations and irregularities. It is the irregular features that make contact with the true geometric counterpart. In some instances where a feature is not well defined, the part might have to be adjusted in order to achieve the best fit. Datum features may be a point, line, surface, axis, center line, median plane, etc.

Datums are specified to convey the design intent clearly to all that read the drawing. Before datums were specified, assumptions were made about the intent of the design. Today, datums are specified for all parts within a design based on the three-plane concept for both circular an noncircular parts. The three-plane concept provides a solid repeatable orientation.

Datum features are identified with a datum feature symbol or datum target symbol. Letters of the alphabet are used to identify the datum features. Datum features may also be only part of a surface, axis, center plane,

etc. If so, the designer will indicate the partial feature with a chain line and required dimensioning for location and length or area of the partial datum. Datums are located either with basic or toleranced dimensions.

Features of actual mating size and patterns of features may also be specified as datum features. These features are permitted actual mating size variation, therefore requiring adjustable gaging to determine the datum. The gaging provides a true geometric counterpart for dimension origins and feature relationship dimensioning.

Chapter 3 Evaluation

1. The 1994 ASME Standard requires that datums are
_____.

2. Datums are theoretically exact and are used for _____ _____ origins and part _____ .

3. List those features of parts that can be used as datums:

_____   _____

_____   _____

_____   _____

4. Is datum reference letter alphabetical order in feature control frames important? _____

5. How many points of contact minimum are required for a primary datum plane? _____

6. The planes of a datum reference frame are assumed to be at _____ degrees basic to each other.

7. Datum target areas are identified with a phantom line _____ with crosshatch lines inside.

8. The upper half of the datum target symbol contains the area _____ when the symbol is attached to a datum target area.

9. Datum targets may be _____ with adjustable gages, pins, collets, etc.

10. Datums of size are features associated with a dimension and _____.

11. Datums of size are _____ with adjustable gages, pins, collets, etc.

12. Datums of size also need additional consideration when the _____ are specified with them.

13. Datums are specified on drawings to _____ a clear intent of the design.

14. The minimum point contact required in the three-plane concept is to eliminate part _____.

15. When a datum target area is specified, a _____ nose pin the size of the specified area is required.

FEATURE CONTROL FRAMES

Introduction

This chapter deals specifically with feature control frames. A feature control frame is a rectangular box with many compartments. These compartments contain the symbols, tolerances, and datum reference letters discussed in Chapters 2 and 3. Where applicable, the tolerance is preceded by the diameter symbol and followed by a material condition symbol. The datum reference letters may also be followed by a material condition symbol. The symbols, when combined in a specific sequence in the feature control frame, provide a specific control instruction for a feature or group of features to which it is attached. The contents of feature control frames must always be specified in a standard arrangement. Each feature control frame means one setup for manufacturing or inspection. Feature control frames may be single, combined, or composite.

Symbol and Definition

FEATURE CONTROL FRAME

The feature control frame provides a specified control for single or multiple features. The rectangular frame is constructed as required by the designer to control one specific feature or group of features. A feature control frame must contain at least a geometric characteristic symbol and a tolerance value. Feature control frames are read from left to right and line by line in the case of composite control frames. A feature control frame may contain the following symbols and tolerance:

Geometric characteristics
Diameter symbol
Tolerance

Figure 4-1

Tolerance modifier
Datum reference letter(s)
Datum modifier

Figure 4-1 shows an example of a feature control frame that was specified to control the position of a feature or group of features. The first symbol in a feature control frame is the geometric characteristic symbol.

Attachment

INTRODUCTION

Feature control frames may be attached to features in various ways. The method of attachment picked by the designer determines the effect of the control specified for that feature or group of features. Feature control frames may be attached to a surface, axis, or center line. With each method of attachment, the feature control is limited to only that portion of the part or feature to which the feature control frame is attached. For example, if a feature control frame is attached to a surface extension line, then only that surface is controlled.

SURFACE

Feature control frames that control surfaces either control the entire surface or just the surface indicated. For example, round features are controlled all around. The control applies all around because usually it is too difficult to orient the part for control on one side.

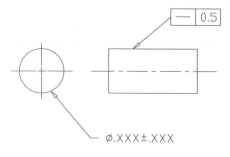

Figure 4-2

Generally, a designer will want the same control to apply all around the part. In Figure 4-2, an example of how a feature control frame is attached to the surface of a circular part is shown.

AXIS OR CENTER LINE

Feature control frames associated with round or width-type features are attached to extension lines of that feature. The interpretation means the axis or center line of the feature is controlled with no regard for the feature surface. The designer is specifying the required control so that fasteners will pass through parts or so that parts will mate with each other. The specification of such callouts means the axis or center line orientation is the critical concern rather than the sides of the feature. Examples of how an axis or center line is controlled are shown in Figure 4-3. (**Note:** Chapter 6 has many examples of feature control frame attachment.)

The feature control frame is usually attached to the controlled feature with one of four methods. These methods are as follows with an example in Figure 4-4:

a) The feature control frame is placed below a dimension pertaining to a feature. The leader is from the dimension.

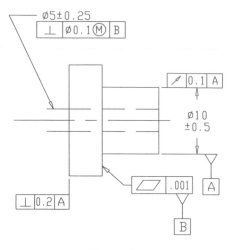

Figure 4-4

b) A leader from the feature control frame to the controlled feature.
c) A side or end of the feature control frame is attached to an extension line from the feature. The feature surface must be a plane.
d) By attaching a side or end of the feature control frame to a feature-of-size dimension extension line.

Content

GEOMETRIC CHARACTERISTIC SYMBOL

The feature control frame consists of many compartments that contain information specified by the designer. The first compartment of the frame always contains the geometric characteristic symbol for form, orientation, or position, as in Figure 4-5.

TOLERANCE

The next compartment always contains a tolerance (see Figure 4-6). The tolerance is either a diameter or

Figure 4-3

Figure 4-5

Figure 4-6

a width. If the tolerance is cylindrical, the diameter symbol will precede the specified tolerance. The tolerance is always a total tolerance and not a plus/minus tolerance as with the coordinate method of dimensioning and tolerancing.

DATUM REFERENCE LETTERS

When required, the datum reference letter or letters follow the stated tolerance (see Figure 4-7). These letters are not always required, and the number of letters may vary from one to three, depending on the datum reference frame required. The alphabetical order is insignificant; the order from left to right is what establishes the precedence for the datum reference frame. The first letter identifies the primary datum plane, the second letter identifies the secondary datum plane, and the third letter identifies the tertiary datum plane.

MODIFIERS

Modifiers are required in certain situations, or they may be specified in other cases depending on design requirements. Modifier specification is explained in

Figure 4-7

Figure 4-8

detail in Chapters 5 and 6. The modifier symbols appear in feature control frames, as shown in Figure 4-8.

How to Read Feature Control Frames

LEFT TO RIGHT

Feature control frames are read from left to right. If the feature control frame is a composite or combined symbol, you must read the first line and perform that requirement. Proceed with the next line or lines performing each requirement in succession. Feature control frames do not specify an order of processing, but state the final requirement for the design. Figure 4-9 illustrates how to read two feature control frames. Note that a feature control frame is required for each feature or group of features to be controlled. (**Note:** There are conditions that require modifiers to be specified, whereas in other cases they are implied. See Chapter 5, Rule 2 for more information.)

Datum Reference Letter Precedence

THE THREE-PLANE CONCEPT

Datum reference letters are arranged in the feature control frame for a specific orientation of the part. The datum reference letters specify the three-plane concept. The first datum reference letter in the feature control frame determines the primary datum, the second letter identifies the secondary datum, and the third letter identifies the tertiary datum. Feature control does not always require three datum reference letters, as shown in Figure 4-9. The designer will specify the number of letters (datum references) required for proper part orientation and feature control. The datum reference letters are also specified in the feature control frame from left to right in their order of precedence; see Figure 4-10.

Figure 4-9

Figure 4-10

TWO DATUM FEATURES

There are some designs that require the identification of two features as datum features to establish a single datum plane, axis, etc. In such cases, each feature is identified with a datum reference letter. (See Figure 4-11.) These two letters then share the same compartment in the feature control frame and are separated by a dash.

As illustrated in Figure 4-11, the center portion of this part is controlled in relation to datum A–B. (The .500 and 1.000 diameters establish datum A–B through the part.) These two features establish a single datum axis through the part.

Figure 4-11

There is one other situation where two letters appear in the same compartment. That situation is when all of the letters of the alphabet are used on one design and there are still more datum features to be identified. In this case, a double letter is used to identify a single datum feature. An example of such a feature control frame is illustrated in Figure 4-12. This application is not found frequently. Usually designs are not complex enough to require such identification.

Figure 4-12

Figure 4-14

Types

INTRODUCTION

Feature control frames may be combined in many different ways. Regardless of the combinations used, there are only three types of feature control frames. There are those that provide control of a single characteristic; there are many combinations of combined feature control frames; and there are those interpreted one line at a time. Once the first line of a combined or composite feature control frame is used, then the next line comes into effect. The following sections give some examples of the various combinations that may be created to control the design intent properly.

SINGLE FEATURE CONTROL FRAMES

A single feature control frame only provides one control for a feature or pattern of features. The control may be specified to control a surface or to control a feature. Figure 4-13 illustrates two examples of single feature control frames.

COMBINED FEATURE CONTROL FRAMES

Combined feature control frames are just what their name indicates. There are two or more frames joined together, or there may be a feature control frame and datum reference symbol combined. These symbols are also interpreted one line at a time. For

example, as shown in Figure 4-14, a combined symbol first specifies the feature to be parallel to datum "D" and then refines that surface to a flatness of less tolerance.

Another application could be a profile tolerance. Figure 4-15 illustrates a combined feature control frame that specifies a surface control and then refines that control to any line with a lesser tolerance.

Another type of combined feature control frame is where a feature is given a specified tolerance and then identified as a datum feature. This may be common practice where one feature or group of features must be in a specific relationship to another feature. Figure 4-16 illustrates two different combined feature control frames.

COMPOSITE FEATURE CONTROL FRAMES

Composite feature control frames are specified to provide a maximum tolerance to orient or position a feature and then refine that feature to a tighter tolerance. A composite feature control frame will only con-

Figure 4-15

Figure 4-13

Figure 4-16

Figure 4-17

tain one geometric characteristic symbol; a combined feature control frame may contain two different symbols.

With composite controls, the first line of the feature control frame is considered to be the largest orientation or location tolerance allowable for the feature(s). Then, once the feature is within this tolerance, it is refined to a closer tolerance to control it to ensure assembly with a mating part. An example of a composite feature control frame is shown in Figure 4-17. This composite control will be explained in depth in Chapter 8, and it is most frequently specified with location tolerances.

Summary

The feature control frame is specified for each feature or group of features. These feature control frames provide one instruction concerning the form, orientation, or position of features. This means that each feature control frame requires a setup for manufacturing and inspection. The feature control frame contains the information for proper part orientation in relation to the specified datums. The datum reference letters in the feature control frame denote the datum precedence in relation to the three-plane concept.

Feature control frames may be attached to features in one of four ways. The selected method of attachment by the designer controls how features are controlled. The feature control frame may be constructed as single, combined, or composite. Regardless of the type of feature control frame, they all are read from left to right and one line at a time. When one line is read and applied, that line is finished. The information is only used once for a feature.

Chapter 4 Evaluation

1. Feature control frames are rectangular boxes that contain specific information to _____ a feature or group of features.

2. Feature control frames may be single, _____ , or composite.

3. Feature control frames may be attached to a _____ axis or center line.

4. The first symbol in a feature control frame is a _____ _____.

5. The feature control frame must always contain a specified _____.

6. Feature control frames are always read from _____ to _____.

7. When datum reference letters are specified in a feature control, the first letter always identifies the _____ datum.

8. When reading a composite feature control frame, you always read one _____ of it at a time.

9. Feature control frames consist of a various number of _____ that contain symbols.

10. Is the alphabetical order of the datum reference letters important in the feature control frame? _____

GENERAL RULES

Introduction

Figure 5-1

Geometric Dimensioning and Tolerancing (ASME Y14.5M-1994), like most other standards, contains specific rules. This standard identifies three rules, but in fact, there are four general rules that apply in various situations. These rules are provided to control some general situations and to provide a common foundation to apply and interpret GD&T. The rules provide a means to control these situations with only one interpretation of industrial drawings throughout the world. The rules pertain to feature actual mating size and form tolerances, the specification of modifiers, and the origin of datums.

Rule Two (a): RFS may be specified on the drawing for positional tolerance with respect to the individual tolerance, datum reference, or both, as applicable (see Figure 5-1).

Rule Three: For screw threads, splines, and gears, the tolerance and datum reference originate from the pitch cylinder axis.

Rule Four: A virtual condition exists for features of size and datum features of size.

Overview

Most standards have limited "unposted" rules that must be observed at certain times, like driving, for example. GD&T also has "unposted" rules that must be observed by designers and those who interpret drawings. We know to drive on the right-hand side of the highway in the United States or that we must drive with lights on at times of impaired vision. GD&T also is a standard with "unposted" rules that we must learn and apply when required. These rules, in brief, are as follows.

Rule One: When only a tolerance of size is specified, that tolerance controls both size and form.

Rule Two: Regardless of feature size (RFS) is implied for all geometric tolerances. Maximum material condition or least material condition must be specified on the drawing where it is applicable. **Note:** Circular runout, total runout, concentricity, and symmetry are applicable only on an RFS basis and cannot be modified to MMC or LMC.

Rule One

Rule One applies to all features controlled with only plus/minus tolerances. Rule One states, "where only a tolerance of size is specified, the limits of size of an individual feature prescribe the extent to which variations in its geometric form, as well as size, are allowed." In such cases, the individual feature size limits control both the amount of variation in actual mating size and the form. The actual local size of an individual feature at any cross-section shall be within the specified size and plus or minus tolerance. Although the feature must meet size requirements at any cross-section, the form must also be within these size limits.

The feature surface or surfaces may not exceed the limits of size. These size limits become the boundary of perfect form. For features like pins, the boundary is the features MMC and no variation beyond this actual local size is permitted. For internal features, the MMC is the nominal size minus the negative size tolerance. Rule One simply permits variation in a features form

Ø.500 +.015 -.003

Ø.520 +.015 -.003

Figure 5-2

MMC Ø.515

LMC Ø.497

LMC Ø.535

LMC Ø.497

Figure 5-4

based on its produced actual mating size. Figure 5-2 illustrates a drawing specification.

When the feature varies or departs from MMC toward LMC, its form is allowed to vary from perfect. Parts are given tolerance to allow for variation, because the closer to perfect a feature is produced, the more expensive the parts are. Therefore, if variation is allowable, the designer will allow as much as possible based on the design requirements. Features then may wave, bow, taper, step, etc., an amount equal to their departure from MMC. Figure 5-3 is an illustration of feature variation.

When the feature departs from MMC to LMC, there is no requirement for perfect form. The feature is allowed to vary the full limit of the size tolerance. Any variation in the form is acceptable. The variation bow,

Ø.515   Ø.497

LMC Ø.535

MMC Ø.517

Figure 5-3

for example, cannot exceed the MMC boundaries. Figure 5-4 shows an example of how the two mating parts may be produced.

PERFECT FORM NOT REQUIRED

There are certain conditions when Rule One is not desired or does not apply. Designers may wish to permit a feature to vary beyond the boundary of perfect form at MMC. In such designs they may add a note, "PERFECT FORM AT MMC NOT REQD." This boundary of perfect form may also be violated by Rule Four, the datum/virtual condition rule. This rule is discussed in this chapter and in detail in Chapter 7.

Rule One does not control the geometric form of commercial stock and parts subject to free-state variation. Commercial stock would include bar, sheet, tubing, or structural shapes, or any products produced to established industry or government standards. Parts subject to free-state variation include rubber and plastic, for example.

FEATURE RELATIONSHIP

Rule One does not control the interrelation of features; it only applies to the individual feature of size. If interrelationship is to be controlled, other geometric controls must be specified. Figure 5-5 illustrates a piece of tubing; the inside diameter is an individual feature completely separate from the outside diameter.

Figure 5-5

Figure 5-6

Rule Two

Rule Two is primarily a designer's rule. The rule states, "Regardless of Feature Size (RFS) applies, with respect to the individual tolerance, datum reference, or both, where no modifying symbol is specified. Maximum material condition (MMC) or least material condition (LMC) must be specified on the drawing where it is applicable. **Note:** circular runout, total runout, concentricity and symmetry are applicable only on an RFS basis and cannot be modified to MMC or LMC." The designer must specify MMC or LMC for features subject to size variation. Features subject to size variation include those being positioned as well as any datum features subject to size variations. An example of such a feature control frame is shown in Figure 5-6. Various other feature control frames are used in Chapter 8.

Rule Three

Rule Three applies to all screw threads, gears, and splines and states that "for each tolerance of orientation or position and datum reference specified for screw threads applies to the axis of the thread derived from the pitch cylinder, for gears and splines the MAJOR DIA., PITCH DIA., or MINOR DIA. must be specified." The designer will specify one of these abbreviations beneath the feature control frame. When there is an exception to the pitch cylinder diameter for screw threads, the designer may specify major dia., pitch dia.,

or minor dia. In Figure 5-7 are illustrations of a feature control frame and datum feature symbol with such notation.

Rule Four

Rule Four, or the datum/virtual condition rule, applies to datum features subject to size variation. The rule states, "depending on whether it is used as a primary, secondary, or tertiary datum, a virtual condition exists for a datum feature of size where its axis or center line is controlled by a geometrical tolerance."

Figure 5-7

Figure 5-10

Figure 5-8

Virtual condition is the worst acceptable condition of a feature. Datum features of size apply at their virtual condition even though they are referenced in the feature control frame at MMC; see Figure 5-8. When the designer does not intend for virtual condition to apply for a primary datum, the feature control frame is associated with the size dimension or is attached to an extension of the dimension line.

For secondary or tertiary datum features of size in the same datum reference frame, the size of the simu-

lated datum is the virtual condition of the datum feature. In Figure 5-9, datum "Y," the axis of the actual 2.250 diameter, is the simulated datum. Datum "Z" is simulated with a virtual condition width perpendicular to datum plane "X." The center plane of this simulated datum is aligned with datum axis "Y."

Another method of specifying tolerance where the design requirements disallow virtual condition is to specify an appropriate geometric control with a zero tolerance at MMC for features of size. Figure 5-10 illustrates such a requirement.

Summary

The GD&T rules are critical to understand if drawings are to be interpreted properly. These rules provide guidelines for both the designer and the interpreter of drawings dimensioned with GD&T. The rules must become part of the interpretation process. It becomes very easy to get involved in the effect of the rules.

Parts may be accepted or rejected incorrectly if some of these rules are ignored. Some rules control form and interrelationship of features; some control the tolerances of form, orientation, and position. One rule controls the datums and virtual conditions of features. These rules must be summarized and remembered in order to interpret drawings properly.

These rules will be applied and referred to in the following chapters. Various examples will aid in remembering their application and effect. The rule number is not always referred to in later chapters when a rule controls a specific application.

Figure 5-9

Chapter 5 Evaluation

1. Rule One controls both _____ and _____ of features.

2. Rule One only controls _____ features, not the interrelationship of features.

3. Rule Two specifies that _____ or _____ must be specified in the feature control frame where it is applicable.

4. Rule One specifies a boundary of perfect form at ____ _____ for features of size.

5. When virtual condition is disallowed, _____ positional tolerance may be specified.

6. _____ applies, with respect to individual tolerance, datum reference, or both, where no modifying symbol is specified.

7. Tolerances and datum references originate from the _____ diameter of screw threads.

8. For gears and splines, the designer will specify MINOR, MAJOR, or PITCH DIA. _____ the feature control frame.

9. The basic rules of GD&T provide a common _____ _____ to apply and interpret GD&T.

10. These rules must be memorized because they are ____ _____ specified on the drawing.

FORM AND ORIENTATION TOLERANCES

Introduction

Geometric dimensioning and tolerancing provides the designer with a means to define mating parts completely. The portion of the GD&T standard that provides methods of controlling part features is the form and orientation controls. These controls are specified for features critical to function and interchangeability where tolerances of size and location do not provide adequate control. When form and orientation tolerances are specified, the tolerance for some features may increase with the use of modifiers.

Remember, GD&T is not a replacement for the coordinate system. It is used in conjunction with it to describe completely design requirements. Also, remember that now you have two tolerances to work with. The first, and it should be considered first, is the plus or minus size tolerance. Then geometric tolerances of form, orientation, and location must be considered. You will learn that these tolerances are interrelated in certain applications.

Form tolerances are most frequently applied to single features or portions of a feature. These controls are specified without a datum reference because the features are not controlled in relation to another feature. Orientation tolerances control features in relation to one another, therefore, a datum reference is required. All dimensions and tolerances apply at free state unless otherwise specified.

Straightness

SYMBOL

—

Figure 6-1

DEFINITION

Straightness is the condition where one line element of a surface or axis is in a straight line.

TOLERANCE

Straightness tolerance provides a zone in which a surface element or axis must lie. For surface control, the tolerance is implied regardless of feature size (RFS) because the straightness tolerance controls line elements that have no size. For axis control, the tolerance is implied RFS if a modifier is not specified. The tolerance is applied in a view of the drawing where the controlled elements, surface or axis, are shown as a straight line. Each line element or axis must lie within the limits of size for the feature. The tolerance zone may be a width or diameter. The feature must be within the stated size limits at each cross-sectional measurement.

SURFACE CONTROL

When a surface is to be controlled, the feature control frame is attached to the surface with a leader or extension line. In the case of cylindrical features, the entire surface must be checked. First, all elements of the surface must be within the specified size tolerance, and then within the limits of the straightness tolerance zone, which is also within the size limits (tolerance). In Figure 6-2 are examples of how the straightness control may be applied to flat and cylindrical features.

TOLERANCE INTERPRETATION

The tolerance zone is a space between two parallel straight lines that may make contact with the surface of the feature. The tolerance zone for both flat and cylindrical features is applied along the entire surface. This surface may be measured or verified with a dial indicator or any other digital readout. The only concern here is how much of the surface has to be verified at one line element at a time. Enough line elements must be verified to ensure that the part or feature is within

Figure 6-4

Figure 6-2

feature the tolerance only applies to the surface indicated, as in Figure 6-2. The tolerance will only apply to the top surface of the part.

AXIS CONTROL

design requirement. There is no hard and fast rule as to how many measurements to make.

The feature surface may take any shape such as barreling, waving, concaveness, etc., as long as it meets the size requirements, and then falls within the specified form tolerance at full indicator movement (FIM). In Figure 6-3 are illustrations of how the features may appear. Straightness control specified for a flat part or surface would have a tolerance zone like those shown in Figure 6-3. The major difference is that for round features the tolerance applies all around, and for a flat

To control an axis, the feature control frame is specified with the diametral feature size (Figure 6-4). The control is specified in a view where the axis is shown as a straight line. This type of attachment means that the axis has a specified diametral zone to lie in.

TOLERANCE INTERPRETATION

The axis of the feature may take any form as long as it stays within the diametral zone. If modifiers are not specified, then the tolerance is implied as RFS. If applicable, maximum material condition (MMC) may be specified for this form control. When MMC is specified, the collective effect of the feature actual mating size and the straightness tolerance may result in a virtual condition (Figure 6-5); in other words, the boundary of perfect form may be exceeded to the limit of the stated tolerance.

Because the feature in Figure 6-4 is a feature of size, the modifier principles do apply if desired by the designer. The feature control frame for the part in Figure 6-4 could have contained a modified tolerance, $\varnothing0.5$ Ⓜ. If the tolerance was specified in this manner, the toler-

Figure 6-5

Figure 6-3

FEATURE SIZE	TOLERANCE ZONE
12.3	0.5
12.4	0.6
12.5	0.7
12.6	0.8
12.7	0.9

Figure 6-6

ance zone diameter in which the axis must lie could vary, based on the actual diameter of the part. When the part is produced at 12.7 mm, the tolerance zone is Ø0.5 mm as specified in the feature control frame. The diametrical tolerance zone could increase to Ø0.9 mm when the part is at the smallest actual mating size or least material condition (LMC). See the tolerance table in Figure 6-6. This same control may be specified for noncylindrical parts. If straightness is specified for noncylindrical features, the diameter symbol will not precede the stated tolerance in the feature control frame.

Flatness

SYMBOL

∠7

Figure 6-7

DEFINITION

Flatness is the condition of a surface where all elements are in one plane.

TOLERANCE

Flatness tolerance provides a zone of a specified thickness defined by two parallel planes in which the surface must lie. The specified tolerance in the feature control frame is implied as RFS. MMC does not apply to flatness control because only surface area is controlled and area does not have size. The designer will specify the flatness control in a drawing view where the controlled surface elements are shown as a straight line. Then each surface element must lie within the stated form tolerance zone. The form tolerance zone must be contained within the limits of the feature size. Under no circumstances is the feature to exceed the specified limits of size or perfect form at MMC. The feature control frame may be attached to the feature with an extension line of the controlled surface, or attached with a leader pointed to the controlled surface.

APPLICATION

The example in Figure 6-8 illustrates a proper specification of the flatness form control. The feature is given a size and tolerance that must not be exceeded. Then an additional form control tolerance is applied to the controlled surface. The form tolerance *does not* allow the feature to exceed the specified actual mating size requirements.

TOLERANCE INTERPRETATION

Flatness tolerance is the specified distance between two parallel planes of which the upper limit plane must contact the actual feature surface (see Figure 6-9.) The other plane then should be the stated tolerance from the first and below all surface area irregularities. The actual surface may be verified with a dial indicator. The indicator should be zeroed for the highest or lowest point on the surface. Then the surface is to be checked sufficiently in all directions to ensure that it is within the specified tolerance. The readings obtained are FIM and must not exceed the stated FIM tolerance in the feature control.

Figure 6-8

Figure 6-9

Figure 6-11

Circularity

SYMBOL

Figure 6-10

DEFINITION

Circularity is roundness. Circularity is a condition of a cylindrical surface, other than a sphere, at any cross-sectional measurement where all points of the surface are perpendicular to and equal distance from a common axis during one complete revolution of the feature. Circularity of a sphere is a condition where all points of the surface intersected by any plane passing through a common center are equal distance from that center.

TOLERANCE

Circularity tolerance provides a circular zone in which all points of a cross-section or slice of the surface must lie. The tolerance zone is two concentric circles that are the stated tolerance apart. The specified tolerance is implied to be RFS and FIM. Because circularity is a surface control, the modifier principles do not apply. The feature control frame is usually specified in the end view. The tolerance zone must be within the size limits of the feature. All surface elements must be within the boundary of perfect form at MMC. This tolerance is not associated with a datum. The surface is controlled or compared to itself, the axis; therefore a datum is not required.

APPLICATION

The circulatory tolerance is applied to compare the circular elements or slices of cylindrical features. Figure

6-11 illustrates the proper application of a circularity tolerance. This tolerance may be specified for any cylindrical feature such as cones, spheres, or cylinders that only require line control around the feature. Circularity may also be specified for internal features that are circular in cross-section.

TOLERANCE INTERPRETATION

The circularity tolerance is the space between two concentric circles that is the stated tolerance apart. Figure 6-12 illustrates how the tolerance applies to an external feature. (**Note:** The difference between the two diametral measurements is 0.2, or twice the specified tolerance.) Circularity tolerance is a *radial* tolerance. The larger circle must make contact with the actual surface of the controlled external feature. Then the smaller circle would have to be the stated tolerance away from the larger one or the same as the feature's

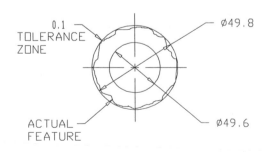

Figure 6-12

smallest permissible diameter. The opposite is true for internal circular features. This tolerance zone is applicable to each cross-sectional element of the feature. The tolerance zone must be perpendicular to the controlled feature axis. All elements of the controlled feature must be within the specified size limits.

Controlled features may be verified with several instruments. The primary concern, however, is how the feature is measured. For example, if a V-block is used, the measurement may include unwanted variables such as lobing, out-of-straightness, the composite effects of a diametrical reading, etc. If possible, the measurements should be made in relation to the axis, because the specified tolerance is on the radius. In this way, all readings on the indicator will be radial, as the tolerance is intended. Regardless of the method of measurement, sufficient measurements must be made to ensure feature acceptance. The specified tolerance is implied FIM for each circular element of the controlled feature.

Cylindricity

SYMBOL

Figure 6-14

Figure 6-13

DEFINITION

Cylindricity is the condition of an entire feature surface during one revolution in which all surface points are an equal distance from a common axis.

TOLERANCE

Cylindricity tolerance provides a zone bounded by two concentric cylinders in which the controlled surface must lie. Cylindricity tolerance is a *radial* tolerance. The specified tolerance is implied to be RFS in relationship to the feature axis. Because the feature is compared to itself, a datum reference is not required. The tolerance is specified in either drawing view. The feature control frame is attached to the feature with a leader. The cylindricity tolerance zone must be within the limits of feature size. At MMC

the feature must have perfect form. Cylindricity tolerance may be applied to internal or external features.

APPLICATION

Cylindricity tolerance is specified in addition to the specified feature size and tolerance. This cylindricity tolerance remains the same size for all possible actual mating sizes and is not an addition to the feature size or tolerance. The cylindricity tolerance is a composite control. It controls feature circularity, straightness, and taper. When a cylindricity tolerance is applied, a datum reference is not required because the feature is compared to itself, the axis. In Figure 6-14, an example illustrates how cylindricity is applied.

TOLERANCE INTERPRETATION

Cylindricity tolerance is the space between two concentric cylinders the stated tolerance apart. The largest cylinder must make contact with the actual surface of the controlled external feature. Then the smaller cylinder is the stated tolerance away from the large cylinder. Neither of the cylinders is to exceed the feature size limits. When cylindricity tolerance is specified for an internal feature, the cylinders establishing the tolerance zone would be opposite. This tolerance zone controls all points of the surface at one time. The tolerance zone is an equal distance from the controlled feature axis (radial measurement) for the length of the feature.

The controlled feature may be verified with several different measuring devices. This control, however, is composite, and the method of measurement is important. If gaging is not used, then an inspection method

Figure 6-15

must be used that will detect excessive taper, out-of-straightness, and out-of-roundness in relation to the feature axis. The measuring device must not indicate more than the stated tolerance. Figure 6-15 shows an example of an external feature and the specified tolerance zone.

Perpendicularity

SYMBOL   ⊥

Figure 6-16

DEFINITION

Perpendicularity is the condition of an entire surface, plane, or axis at a right angle to a datum plane or axis.

TOLERANCE

Perpendicularity tolerance provides a zone defined by two parallel planes, two parallel lines, or a cylinder parallel to a datum. The controlled feature surface, plane, or axis must lie within the specified tolerance zone. Perpendicularity is an orientation control; therefore, a datum reference is required, and the orientation tolerance is implied RFS if not modified to LMC or MMC. Remember, only features of size can be modi-

fied, not surface or line features. The feature control frame is specified in drawing a view where the relationship of features is shown.

APPLICATION

Perpendicularity tolerance is specified for designs that require one feature to be perpendicular to another; therefore, a datum reference is required. This datum error is not considered when measuring the controlled feature. Perpendicularity controls all surface error including flatness and angularity. This control is specified in addition to actual mating size requirements and may be specified in four different methods. A perpendicularity tolerance may be specified for a surface perpendicular to a datum plane, an axis perpendicular to an axis, an axis perpendicular to a datum plane, or line elements of a surface perpendicular to a datum axis. Each of these applications will be explained in detail here.

FEATURE SURFACE PERPENDICULAR TO A DATUM PLANE

This application is the most frequent application for perpendicularity; Figure 6-17 illustrates this application of perpendicularity. The feature control frame is specified in a drawing view where the relationship

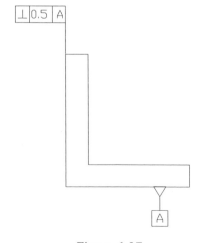

Figure 6-17

between features appears. The feature control frame may be attached to the feature with an extension line or leader.

Tolerance This perpendicularity tolerance provides a zone defined by two parallel planes that are the distance of the specified tolerance apart. The tolerance is implied RFS here, because only surfaces are controlled. The tolerance zone must be within the limits of the feature size. All elements of the controlled feature must lie within the perpendicularity tolerance zone.

Tolerance Interpretation The controlled feature must first meet the feature size requirements, and then the geometric orientation tolerance. The datum reference feature must be placed in contact with the datum plane. Then the tolerance zone must be established at exactly 90° to the datum. All elements of the controlled surface must lie within the parallel planes of the tolerance zone, as in Figure 6-18. (**Note:** Datum error is not additive to the feature being controlled.)

The actual feature surface may be verified in a number of ways. A simple check can be made with a dial indicator or coordinate measuring machine. The datum feature must be placed in contact with the datum plane. Then with a measuring instrument, make contact with the controlled surface. Zero the measuring device on a high point and continue to measure the entire surface. The FIM is not to exceed the tolerance specified in the feature control frame. Enough of the surface must be measured to ensure that design requirements are met.

Figure 6-19

FEATURE AXIS PERPENDICULAR TO A DATUM AXIS

Perpendicularity specified in this manner is not a common application for this control. However, this use of perpendicularity does provide the required control for designs where a pin or hole must be at a 90° angle to the axis of another cylindrical part or feature. Figure 6-19 illustrates an application of this orientation control. The feature control frame is specified in a drawing view where the relationship between the datum axis and the controlled feature are shown. The feature control frame is attached to the controlled feature with the feature size call out.

Tolerance The perpendicularity tolerance when specified in this manner provides a tolerance zone for the axis to lie within. The tolerance zone may be a width between two lines or a cylinder. The zone is a width when the diameter symbol is *not* specified in the feature control frame. The specified tolerance is RFS if the designer does not specify a modifier. In the example in Figure 6-19, the hole is a feature of size; therefore, the modifiers may be specified depending on design requirements. The axis of the feature must be within the boundaries of the tolerance zone if it is a width or diameter at RFS or modified.

Tolerance Interpretation The controlled feature must meet actual mating size requirements; then geometric controls can be considered. The geometric tolerance zone must be established at 90° to the datum axis. The controlled feature axis must then be within the width or cylindrical zone as specified in the feature control frame. The feature axis may take any shape—

Figure 6-18

Figure 6-20

Figure 6-21

bow, wave, angle—as long as it remains within the tolerance zone. The example in Figure 6-20 illustrates a width-type tolerance zone. The tolerance applies to the feature as a width in the drawing view shown. If the diameter symbol is specified, the tolerance zone would be cylindrical, allowing the axis of the hole to lean or move in all directions.

The actual controlled hole may be verified with several different methods. This feature may best be verified by using a gage rather than by measuring the angularity. Direct measurements would provide the required information to make this check. If a gage were used for verification, it would simulate the mating part. This gage would have to simulate acceptance of the worst condition for the controlled feature. The worst condition is the collective effect of actual mating size and geometric tolerances. This collective effect is known as virtual condition. Rule Four controls datum feature "A." Virtual condition is explained and illustrated in Chapter 7.

FEATURE AXIS PERPENDICULAR TO A DATUM PLANE

Specifying perpendicularity to control the relationship between a cylinder and plane in this manner is a common application of perpendicularity. The feature control frame is attached to the controlled cylindrical feature with the feature size callout. This callout is in a drawing view where the relationship between features can be seen. Figure 6-21 illustrates a proper specification. The perpendicularity tolerance is usually specified

in a drawing view where the relationship is clearly shown.

Tolerance A perpendicularity specification like this is usually a refinement of location tolerancing. Each control, location, and perpendicularity will have a tolerance specified. The axis of the controlled feature must lie within the tolerance(s). The tolerance zone for this application is usually diametral. Because the controlled features are features of size, the modifiers may be specified depending on the final assembly requirements. The axis of the actual feature must lie within the boundaries of the tolerance zone.

Tolerance Interpretation The controlled cylindrical feature must meet the feature size requirements and then geometric controls. If the feature is controlled with a location tolerance and then refined with perpendicularity, the feature must meet the location tolerance and then the perpendicularity tolerance. The perpendicularity tolerance must be at a 90° basic angle to the datum plane. The controlled feature axis must lie within the tolerance zone as specified in the feature control frame or as modified based on actual mating size. The actual axis may be bowed, angled, wavy, etc., as long as it remains within the tolerance zone. In Figure 6-22 is an example of how the actual feature may be produced and how the tolerance may vary when the tolerance is modified to MMC.

This cylindrical feature may be verified with various inspection methods. The most effective way to check the feature with complete assurance that it will assemble with the mating part is with a gage. The gage will have to be made at the virtual condition. A gage at virtual condition will accept all parts that are acceptable for assembly. This gage takes into account all possible

Figure 6-24

FEATURE SIZE	TOLERANCE ZONE
24.75	1.
25	0.75
25.25	0.5

Figure 6-22

feature errors such as out-of-perpendicularity along with out-of-location.

LINE ELEMENT PERPENDICULAR TO A DATUM

The designer may specify a surface perpendicular to another datum plane or axis. The feature control frame is attached to the controlled surface with a leader in a drawing view where the surface to be controlled can be seen. Below the feature control frame the phrase EACH RADIAL ELEMENT must be added by the designer. Figure 6-23 shows an example of how this control is specified.

Tolerance This perpendicularity tolerance provides a tolerance zone defined by two parallel lines that are

the stated tolerance apart. This zone must be perpendicular to a datum plane or axis. Then each controlled line element must lie between the two lines of the tolerance zone. This tolerance is always RFS because surface elements are being controlled. The tolerance zone must be within the size limits of the controlled feature.

Tolerance Interpretation The controlled feature must first meet the feature size requirements. Then the perpendicularity tolerance is applied by making contact with the actual part surface line by line. The tolerance zone is the thickness of the specified tolerance and perpendicular to a datum. Each line element of the controlled surface is subject to complying with the specified tolerance. Figure 6-24 illustrates how the tolerance zone will appear for a controlled feature.

This feature may be verified with basic precision measurement devices. The controlled feature must be verified at 90° from the datum axis or plane. A simple check with a dial indicator or coordinate measuring machine along line elements of the controlled surface is all that is required. Adequate measurements must be made to ensure design intent. Each line element of the controlled feature must lie within the specified tolerance between two parallel lines. The surface form may vary from measurement to measurement, but each line element must be within tolerance at FIM and RFS.

Angularity

SYMBOL

Figure 6-25

EACH RADIAL ELEMENT

Figure 6-23

DEFINITION

Angularity is the condition of an axis or plane other than 90° to another datum plane or axis.

TOLERANCE

Angularity tolerance provides a zone defined by two parallel planes that are the stated tolerance apart and at the specified basic angle to the datum reference. The controlled feature surface, plane, or axis must lie within this zone. Angularity is another of the orientation tolerances; therefore, a datum reference is always required. The tolerance may be modified if a feature of size is being controlled. If modifiers are not specified, Rule Two governs the tolerance. The feature control frame is specified in a drawing view where the angular relationship is shown. This relationship must be specified with a basic angle. The datum reference feature irregularities *do not* affect the controlled feature surface or axis.

APPLICATION

A control is specified to control features that are required to be at an angle other than 90° in relation to another feature plane, or axis. Angularity tolerance controls surface, plane, or axis errors within the limits of the tolerance zone. Angularity also controls flatness and straightness. The angularity tolerance is in addition to the feature size tolerance. The tolerance provides a zone that controls feature origin in relation to another datum feature. Figure 6-26 illustrates a surface application.

Figure 6-26

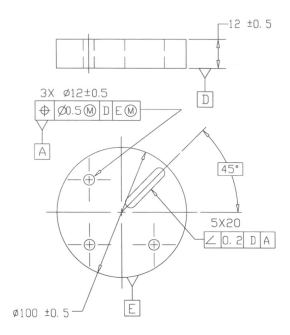

Figure 6-27

When angularity is specified for an internal feature such as a slot or hole, the tolerance only applies in the view and relative to the datums indicated. The feature is not controlled in any other direction. The tolerance is implied RFS, but may be modified depending on the final requirements of the feature. If angularity is specified for a feature of size, it is usually a refinement of a location control. Figure 6-27 illustrates an axis control.

TOLERANCE INTERPRETATION

The controlled feature must meet all other tolerances, and then the orientation control for angularity. The datum reference feature irregularities are not considered when measuring angularity. The tolerance is established at the specified basic angle to the datum. The outer plane of the tolerance zone contacts the highest point(s) on the controlled surface. The inner plane is at the specified tolerance from the first. The controlled surface may take any form through this tolerance zone. The beginning of the angle must be within the length requirement of the part, and may be curved, bowed, twisted, or at a different angle. Figure 6-28 illustrates how a toler-

Figure 6-28

ance zone for a surface is applied. The surface may be verified with a dial indicator or coordinate measuring machine.

The control of an axis or plane is slightly different than the control of a surface. The major difference is the tolerance zone. The tolerance only applies in the view that it is specified in and relative to the indicated datums. The actual tolerance zone, then, is a slice the thickness of the specified tolerance that passes through the controlled feature. When viewing the feature as shown in the drawing view with the orientation control, the end of the tolerance would be seen. This zone extends the total length of the part. Figure 6-29 illustrates the tolerance zone for a slot or hole.

Angularity may be verified with several methods. If a modifier were specified, then a gage would be the most effective way to verify correct feature angularity. If the feature is toleranced at RFS, an expandable gage pin, dial indicator, or coordinate

measuring machine might be used to determine correct angularity.

Parallelism

SYMBOL

Figure 6-30

DEFINITION

Parallelism is the condition of a surface, center plane, or axis that is an equal distance at all points from a datum plane or axis.

TOLERANCE

Parallelism tolerance provides a zone defined by two parallel planes, lines, or a cylinder parallel to a datum plane or axis within which the surface elements or axis of the controlled feature must lie. The specified tolerance is implied RFS by Rule Two. The designer may specify modifiers for features of size based on design requirements for final assembly. The feature control frame is specified in a drawing view where the parallel relationship is shown. This feature control frame must contain a datum reference letter, because parallelism is an orientation tolerance.

APPLICATION

Parallelism tolerance is specified for designs where a plane, surface, or axis must be controlled for parallelism in addition to the feature size tolerance. A datum reference is required with this orientation control. Parallelism also controls flatness of a surface and straightness of an axis within the limits of the tolerance. When parallelism is specified for a surface, the controlled surface must be within the specified size limits. Parallelism orientation may be specified to control three different conditions. The various applications are a surface parallel to another surface, a cylinder parallel to a surface, and a cylinder parallel

Figure 6-29

Figure 6-31

Figure 6-32

to a cylinder. These various applications are discussed here in detail.

SURFACE PARALLEL TO ANOTHER SURFACE

This is the most frequent application of the parallelism orientation control. Figure 6-31 illustrates an application. The feature control frame is attached to the controlled feature with an extension line in a drawing view where the relationship is shown.

Tolerance The tolerance for the controlled feature is the space between two parallel planes that is the specified tolerance apart and parallel to the datum plane. This zone must be within the limits of feature size and is implied RFS according to Rule Two. All elements of the controlled surface must be contained within the tolerance zone.

Tolerance Interpretation The controlled feature must first meet the actual mating size limit requirements, and then the geometric orientation tolerance. The datum reference feature must contact the simulated datum plane. Any irregularities in the datum reference surface are *not* additive to the controlled feature. The outer tolerance zone plane must be established from the outermost surface elements and must be parallel to the datum plane. The inner plane is then the specified orientation tolerance apart from the outer plane. All controlled surface elements must be within the limits of actual mating size and the parallelism tolerance zone. Figure 6-32 illustrates how the toler-

ance zone would apply to the actual mating size of a feature.

The surface may be verified with the basic instruments of inspection, such as a coordinate measuring machine or dial indicator. The datum surface *must* have the minimum three-point contact with the datum plane. The controlled surface must be measured in many directions and across the entire surface until it is ensured of acceptance.

TANGENT PLANE

When it is acceptable to control a feature surface by contacting points of the surface, the tangent plane symbol is specified in the feature control frame following the stated tolerance. Figure 6-33 illustrates this specification.

Tolerance Interpretation The controlled surface of the feature must lie within two parallel planes 0.5 apart. A plane contacting the high points of the surface must

Figure 6-33

Figure 6-34

Figure 6-36

lie within this established zone. Figure 6-34 illustrates this specification.

CYLINDER PARALLEL TO ANOTHER SURFACE

This is a practice that may be specified when a cylindrical feature is to be controlled in only one direction. Figure 6-35 illustrates an application of this orientation control. Here the controlled feature is only controlled in the direction shown in the drawing view that contains the feature control frame. The controlled feature is usually located with a location tolerance and then refined with parallelism.

Tolerance The tolerance for a controlled feature is a width zone that is the size of the specified tolerance. The axis must lie within two parallel lines that are the length of the controlled feature. The axis may pass through this zone at an angle or the axis may be curved, waved, etc., as long as it is within the width tolerance zone. This tolerance zone must be exactly parallel to the datum plane.

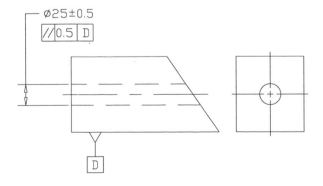

Figure 6-35

Tolerance Interpretation To determine if the controlled feature is acceptable, it must first meet the specified size requirements. Then the part must be placed in contact with the specified datum plane. The simulated datum plane may be a knee or stop on the inspection table or workbench. The part *must* be oriented as shown on the drawing. The cylindrical feature is only controlled in the direction it is shown on the drawing. The tolerance zone for the axis to lie within is determined from the simulated datum plane. This can be achieved with direct measurements. The produced feature must also be measured to determine size and shape. Figure 6-36 shows how the actual feature might appear in relationship to the tolerance zone. (**Note:** The tolerance for this feature may be modified because it is a feature of size. If modified, the tolerance zone would vary depending on actual mating size.)

CYLINDER PARALLEL TO ANOTHER CYLINDER

Specifying a cylinder parallel to another cylinder is one method of controlling cylinders or holes parallel to each other. Depending on the application, this control may provide the required accuracy for final assembly. Figure 6-37 illustrates the application of this parallelism control. The feature control frame is shown in a drawing view where the parallel relationship is required. This parallelism control provides a diametral tolerance zone for the controlled feature.

Tolerance The tolerance zone for the controlled feature may be a cylinder or a width, and modifiers may be specified. The axis of the controlled cylinder must lie within the specified zone for the length of the

Figure 6-37

Figure 6-38

controlled feature. The controlled feature must be within the limits of size and not exceed the boundary of perfect form at MMC.

Tolerance Interpretation The tolerance may be specified in several ways, as just discussed. Here the tolerance will be discussed as a cylindrical zone at MMC; the datum feature is considered at RFS. Then the datum must be established by the largest circumscribed cylinder. The cylinder should be adjustable so that it will make contact with the irregularities of the pin. The axis of this cylinder is the simulated datum used to verify the controlled feature. The adjustable cylinder may be a gage used for verification.

The controlled feature (pin) has the minimum tolerance zone at MMC that must be parallel to the simulated axis of the datum feature. The axis of the controlled feature must lie within that cylindrical zone. The axis may take any form as it passes through the cylindrical tolerance zone. The tolerance zone for the controlled feature is permitted to increase an amount equal to the amount the feature departs from MMC. See Figure 6-38 for an illustration of the tolerance zone and the tolerance increase from MMC to LMC.

Verification of this part is best accomplished with a functional gage. The gage would have holes in it that are used to verify the controlled feature. The hole to simulate the datum must be adjustable to *fit* the produced datum feature. The hole to check the controlled feature would be made at virtual condition.

Another method of specifying the parallelism control

Figure 6-39

for these cylindrical features is to specify both features of size with the MMC modifier. If they were specified in such a manner, the datum feature is controlled by Rule Four, the datum/virtual condition rule. By Rule Four, the datum feature gage hole for simulating the datum would have to be at virtual condition rather than adjustable as it was for RFS. The gage hole for the controlled feature would also be at virtual condition. Figure 6-39 illustrates the part described.

Profile

SYMBOLS

Figure 6-40

DEFINITION

Profile tolerancing is a method of specifying control of deviation from the desired profile along the surface of a feature.

TOLERANCE

Profile tolerances may be specified either as a surface or line profile. The tolerance provides a uniform zone along a desired true profile (a bilateral zone) or a zone defined by a phantom line either inside or outside the basic true profile of the part (a unilateral zone). The surface of the controlled feature must lie within this zone. The specified tolerance is always RFS because only a surface is controlled.

The feature control frame is usually attached with a leader to the surface to be controlled. The feature control frame is specified in a drawing view where the profile is shown. The feature control frame must contain a datum reference and may also have a note added beneath stating "between two points (X and Y)" or "all over."

APPLICATION

Profile tolerance is specified for designs where the surface is to be controlled within a given basic shape. Most frequently, profile tolerances are specified for irregular features that are difficult to control with other form or orientation tolerances; but they may also be specified to control the all around shape of stampings, burned parts, etc. The basic profile of a part is described with basic dimensions, radii, arcs, angles, etc., from the datum. Then the specified profile tolerance controls the amount of deviation in relation to the datum reference(s).

Profile tolerances may be specified to control either a surface or line element of a feature. These two applications are discussed next.

SURFACE PROFILE

Surface profile is a method of specifying a three-dimensional control along the entire surface to be controlled. This control is usually applied to parts

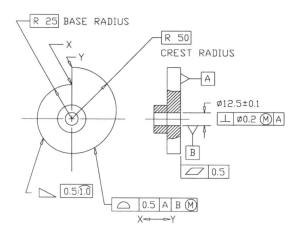

Figure 6-41

having constant cross-section, surfaces of revolution, weldments, forgings, etc., where an "all over" requirement may be desired. Surface control may apply all around. Surface profile may also be specified for coplanar surfaces. Surface profile controls actual mating size as well as shape; therefore, actual mating size and geometric form are verified simultaneously. Figures 6-41 and 6-42 illustrate the proper application of surface profile tolerancing.

Tolerance Surface profile tolerance may be specified as an equal or unequal tolerance on either side of the desired basic profile. If the design requires an unequal application of a bilateral tolerance, a phantom line is included along the basic profile, and the amount of tolerance between the basic profile and the phantom line is specified. Then the balance of

Figure 6-42

BILATERAL TOLERANCE

Figure 6-43

UNILATERAL INSIDE

Figure 6-45

the tolerance is applied to the other side of the basic profile. Phantom lines are also specified when the tolerance is unilateral; all tolerance is applied to one side or the other of the basic profile.

Tolerance Interpretation Profile tolerance for a surface control is implied to be bilateral. If the design requires the tolerance to be unequal or unilateral, phantom lines are added to the drawing view indicating how the tolerance is to be applied. The controlled feature's size and shape are toleranced with the profile tolerance. Surface profile tolerance is a three-dimensional zone. The tolerance applies per-

pendicular to the true profile at all points along the controlled surface. If the tolerance is bilateral, the actual surface of the feature may vary both inside and outside of the true basic profile. If the tolerance is unilateral, then the actual feature may vary only to the inside or the outside of the basic true profile as indicated with a phantom line. Figures 6-43, 6-44, 6-45, and 6-46 illustrate the various tolerance zones permitted by surface profile tolerancing. The tolerance zone is established in relationship to the basic true profile of the feature, *not* the actual feature surface.

The actual part surface may be verified with several techniques. The actual part may be compared to a master part, an overlay may be used, optical comparison can be made, or a dial indicator or coordinate measuring machine can be used. Verification will depend largely on the accuracy required.

UNILATERAL OUTSIDE

Figure 6-44

Figure 6-46

Figure 6-48

Figure 6-47

LINE PROFILE

Line profile tolerancing is a method of specifying a two-dimensional control for a single line element along the true profile of a surface. This control is usually specified for the shape of cross-sections or cutting planes of parts. The control is most frequently specified for manufactured parts for trucks, automobiles, and marine uses (impellers, body parts, propellers, etc.). This control should be used where "blending" is required. Line profile, like surface profile, may be specified between points or all-around. Line profile tolerancing is usually a refinement of some other geometric control, form and size control. The application of line profile is illustrated in Figure 6-47.

Tolerance The tolerance for line profile may be specified as either bilateral or unilateral. The bilateral tolerance is implied to be bilateral by the absence of phantom lines to indicate the tolerance zone. The unilateral tolerance zone is indicated with a phantom line either inside or outside the desired true profile of the controlled part.

Tolerance Interpretation The actual surface must lie within the boundaries of the specified zone. The tolerance applies perpendicular to the line profile at all points along the controlled surface. Line profile control may require datum references; when specified, the feature must be properly oriented when applying the

specified tolerance. The profile tolerance controls both actual mating size and form. The tolerance zone is established in relationship to the true profile of the controlled feature, *not* the actual surface. All points along the controlled surface must lie within the specified tolerance, as shown in Figure 6-48.

Line profile may be verified with the same method used to verify surface profile.

Boundary Control for Noncylindrical Features

Introduction Profile tolerancing may be combined with positional tolerancing where it is necessary to control the boundary of a noncylindrical feature. This application may be specified where it is desirable to establish a tolerance zone to control both the size and shape of a feature (see Figure 6-49).

Tolerance The positional tolerance establishes a theoretical boundary as described by basic dimensions. For internal features, the boundary equals the MMC size of the profile minus the positional tolerance, and the entire feature surface must lie outside the boundary. For an external feature, the boundary equals the MMC size of the profile plus the positional tolerance, and the entire feature surface must lie within the boundary. When this method of tolerancing is specified, the term BOUNDARY is placed beneath the positional tolerance feature control frame.

Tolerance Interpretation Boundary control for a noncylindrical feature is illustrated in Figure 6-50. The profile controlled surface of the feature, all-around,

Figure 6-49

Figure 6-51

Tolerance Composite profile tolerance provides a tolerance for location of a profiled feature as well as the tolerance for form, orientation, and in some instances, the actual mating size of the feature, within the larger profile locating tolerance zone. Each horizontal line of the feature control frame constitutes a separately verifiable requirement. The upper segment is the profile locating control. The lower segment is a refinement control for actual mating size, form, or orientation. Figure 6-51 illustrates a composite profile tolerance feature control frame.

must lie between the boundaries established by the profile tolerance. Then the positional control regulates the surface at MMC. In this illustration no portion of the surface is permitted to lie within the boundary of the MMC contour minus the positional tolerance when positioned in relationship to the datum planes.

COMPOSITE PROFILE

Composite profile is a method of specifying a profile where requirements will permit a feature locating tolerance zone to be larger than the tolerance zone that controls the actual mating size.

Runout

SYMBOLS

Figure 6-52

DEFINITION

Runout is a composite form and location control of permissible error in the desired part surface during a complete revolution of the part around a datum axis.

Figure 6-50

TOLERANCE

Runout tolerances may be specified as either total or circular. The specified tolerance is the deviation permitted in relation to the controlled features axis. The specified tolerance provides a zone between two concentric cylinders for total runout control and between two concentric circles for circular runout control. The surface or all points on a cross-sectional line must lie within the specified tolerance zone. The tolerance is *always* implied to be RFS as measured in relation to the datum axis. The tolerance is specified in a drawing view where the controlled feature(s) is shown. The feature control frame is attached to the controlled feature with a leader or associated with the feature size callout. A datum reference is required.

APPLICATION

Runout tolerance is specified for designs where rotation is involved, such as shafts, pulleys, and bearing surfaces. Runout may also be specified for coaxial features; this control, however, is restrictive for manufacturing because the tolerance is always RFS. Location controls may be a better choice for coaxiality, depending on design requirements. Runout tolerances control the amount of radial deviation for a line or surface of parts that are circular in profile.

TOTAL RUNOUT

Total runout is specified to control feature surfaces that are manufactured with an axis. This control is more stringent than circular runout. It provides a composite control of all surface elements in relation to a datum axis. The controlled surface may be at right angles to or around the datum axis. This control is specified when the composite effect of *all* surface elements together are critical to the final assembly. Figure 6-53 illustrates an application of total runout control.

Tolerance Total runout tolerance is specified in the feature control frame and is implied to be radial, RFS and FIM. The feature control frame is attached directly to the controlled feature. The actual feature must *not* exceed the boundary of perfect form at MMC. The controlled surface must lie within two concentric cylinders that are the stated tolerance apart and an equal distance from the datum axis. The entire actual feature must lie within this tolerance zone simultaneously.

Tolerance Interpretation Total runout tolerance *always* applies at RFS. The tolerance establishes a cylindrical tolerance zone that is the width of the specified tolerance in the feature control frame. This cylinder of tolerance is an equal distance from the datum axis all around. The tolerance is applied simultaneously to all circular and longitudinal elements in one setup during a complete revolution of the controlled feature. The tolerance is composite and cumulative. The feature is controlled for taper, coaxiality, circularity, cylindricity, straightness, angularity, flatness, perpendicularity, and profile. Figure 6-54 illustrates the actual tolerance zone for a controlled feature.

Total runout may be verified by mounting the datum axis in a precision rotational device that will rotate the

Figure 6-53

Figure 6-54

Figure 6-55

Figure 6-56

controlled feature(s) around the datum. The feature may be mounted on a functional diameter or mounted on centers. When the method of measurement is selected, an indicator is set up in contact with the controlled surface and zeroed. The indicator is *not* rezeroed during the measuring operation. The controlled feature is measured parallel to the datum for circular control and perpendicular to the datum for surfaces perpendicular to the datum. The part must be rotated 360° during the measuring operation, and the indicator reading must not exceed the tolerance stated in the feature control frame. Sufficient measurements must be made to satisfy the drawing requirement.

CIRCULAR RUNOUT

Circular runout is specified to control only surface elements of features that are circular in cross-section or surfaces perpendicular to a datum axis. Circular runout is only a line-by-line control of a surface. Each line is completely independent of the other. The tolerance is implied FIM in relation to a feature datum axis. This control is specified when the part function is not critical to rotational speeds. Circular runout is applied in Figure 6-55.

Tolerance The tolerance for circular runout is specified in the feature control frame and is *always* implied RFS. The tolerance is radial and normally implies FIM. The feature control frame is attached to the controlled feature with a leader or with the feature size callout. The actual feature surface must *not* exceed the boundary of perfect form at MMC. Each cross-sectional slice

of the surface must lie within the two concentric circles established by the tolerance zone.

Tolerance Interpretation Circular runout tolerance is a width between two concentric circles the size of the stated tolerance. Each circular element independent of the next must lie within the RFS tolerance zone. The tolerance zone is an equal distance from the datum axis all around. The tolerance controls the cumulative variations in circularity and coaxiality of controlled features around a datum axis. For features perpendicular to a datum axis, the tolerance controls circular elements on the plane. Figure 6-56 illustrates how each circular element is controlled by the specified tolerance.

Circular runout may be verified similar to total runout. The difference is the indicator does not have to be moved along or over the controlled surface. The indicator may be rezeroed for each measurement. The feature must be rotated a complete 360° for each measurement. To ensure acceptability, several independent measurements along the controlled surface should be made.

Unit Control

SYMBOL

There is no symbol for unit control. It is specified as a ratio, usually in the second line of a feature control frame.

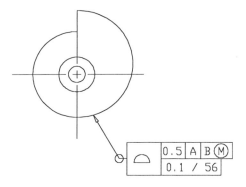

Figure 6-57

DEFINITION

Unit control is a method of controlling feature abruptness in a relatively short length of the controlled feature.

TOLERANCE

The unit control tolerance is usually calculated as a fraction of the form or orientation tolerance. The generally accepted rule of thumb is that the unit tolerance be one-quarter of the stated form or orientation tolerance for any unit of length. Unit control tolerancing is usually an additional tolerance to form and orientation control.

APPLICATION

Depending on design requirements, unit control may be specified for features that are controlled with straightness, flatness, and profile tolerancing. The

Figure 6-58

illustration in Figure 6-57 shows a unit control in conjunction with a profile tolerance. This application specifies a surface profile control of 0.5 mm along the surface. Then units of that surface must also meet the unit control of 0.1 mm per any 56 mm of length.

TOLERANCE INTERPRETATION

The surface profile tolerance is the first consideration in Figure 6-57. If the controlled feature meets the stated requirements, then unit control is applied. Unit control simply means that the profile tolerance of 0.1 mm applies at any unit 56 mm of length along the controlled surface. The units of length may be selected at random or even overlap along the controlled surface. Figure 6-58 illustrates how unit control might apply to a surface controlled with a profile tolerance.

Verification of unit control would depend primarily on the application. Unit control specified with a profile tolerance would have to be verified with the same measuring instruments that are used to verify the profile tolerances. The other applications of unit control can be verified with the instruments used to measure the form or orientation control.

Free-State Variation

Free-state variation is a term used to describe distortion of a part after removal of forces applied during manufacturing. Variation or flexibility is due to thin cross-section, weight, and stress relief. In some cases, parts are required to meet drawing requirements while in free state, whereas others are to be restrained. Restraining forces are those exerted in assembly or functioning of the part. If free-state variation must be controlled, it may be accomplished by specifying tolerances subject to free-state variation or by specifying tolerances on features to be restrained.

Specifying geometric tolerances on features subject to free-state variation Where an individual form or location tolerance is specified for a feature at free state, the maximum allowable tolerance is specified in the feature control frame. This tolerance is followed by

Figure 6-59

any applicable modifiers and then the free-state symbol. Figure 6-59 illustrates such a requirement.

Specifying geometric tolerances on features to be restrained where orientation, runout, or location tolerances are to be verified with the part in a restrained condition, features are selected as datums. Because these datums may be subject to free-state variation, the maximum restraining force must be specified. This force or method of restraint is specified on the drawing in the form of a note. The restraining note may specify a particular type of clamp, bolt and torque value, partial assembly to achieve actual fit, etc.

Summary

The GD&T form and orientation controls are important to understand because they control the required shapes of features. In this summary, each of the controls will be grouped to aid in remembering what and how they control. The groups will address those controls that control lines and those that control surfaces or areas. All of the controls will be compared to the two basic form controls—straightness and flatness. Straightness only controls one line element at a time or the straightness of an axis. Flatness controls all elements or points of a surface simultaneously in all directions. The remaining controls can all be related to straightness and flatness. Consider circularity and cylindricity. Circularity is a line control around features circular in form. It is like straightness around a feature. Cylindricity controls the complete surface area of circular features. Cylindricity is like flatness rolled around a feature.

Perpendicularity may be compared to both straightness and flatness depending on the features being controlled. Perpendicularity may be specified to control four different types of feature-to-datum relationships. When the control is specified for a plane or surface, the control can be related to flatness. When the

perpendicularity control is specified for line element, axis, or median plane, it can be related to straightness.

Angularity is similar to perpendicularity because it controls all angles other than 90°. Some of the angular relationships will be surfaces, and others will be cylinders or noncylindrical features; therefore, angularity can be related to flatness when controlling surfaces. When angularity is specified for cylinders or noncylindrical features, a line or median plane is being controlled like straightness.

Parallelism can also be related to either straightness or flatness depending on the type of feature being controlled. When surfaces are controlled, it is similar to flatness where all elements must be in one plane of given thickness. When features with a median plane or axis are controlled with parallelism, the control is similar again to straightness where a line must be straight within a given tolerance.

Profile tolerances are specified to control either surfaces or lines. These may be features that are irregular where flatness or straightness do not apply. Surface profile tolerance provides a thickness zone around or over the controlled surface like a flatness control would provide. Profile of a line is like straightness where only a line element is compared to the true basic profile. Each line element is considered individually.

The runout tolerances also provide for two types of control, where total runout requires measurement of an entire circular surface in one setup similar to flatness. Circular runout is only a cross-section or slice of a circular feature control similar to straightness. Only one line element at a time is measured and compared to the specified tolerance.

These comparisons provide a means to identify the control provided by each of the geometric form and orientation controls. There are some differences between controls that should be summarized. The primary difference between a runout control and cylindricity is that runout, either total or circular, controls surface-to-axis, whereas cylindricity controls axis-to-axis. Another difference between controls is that of perpendicularity and runout. Perpendicularity is primarily specified to control noncylindrical features. Runout is specified to control cylindrical features.

The final thought or explanation that must be made following all of the above discussion is that the design requirements will always dictate the control to be specified. Parts are not designed for themselves, they are designed for function and relationship in a final assembly.

Chapter 6 Evaluation

1. Flatness is a form control that controls surface elements in all _____ within a specified tolerance.

2. Circularity control applies to feature surfaces during one complete revolution as measured _____ of a surface or axis is in a straight line.

3. Straightness is the condition where one line _____ of a surface or axis is in a straight line.

4. Form and orientation tolerances permit features to vary within the _____ of the tolerance zone.

5. Cylindricity controls the _____ surface of the feature.

6. Per unit control is specified to prevent the continuation of feature _____ or abruptness of the controlled feature.

7. For tolerances of perpendicularity, the zone established by the specified _____ must be within the limits of feature size.

8. The tolerance boundary for a cylindrical feature axis is diametrical when the _____ symbol is specified.

9. Feature control frame _____ determines whether a tolerance is applied to the median plane, center line, or axis of a controlled feature.

10. Form and orientation tolerances are _____ to be RFS.

11. When a form or orientation tolerance is specified for a feature in relation to a datum feature, the datum feature is _____ to be theoretically exact.

12. With the application of GD&T, there are two tolerances allowed: _____ and _____.

13. Angularity is the condition of a surface or _____ at an angle other than 90° from a datum.

14. Parallelism is the condition of a surface or axis an equal _____ at all points from a datum plane or axis.

15. For noncylindrical features, angularity tolerance is a _____ and not an angular tolerance zone.

16. Total runout is always implied _____.

17. A _____ tolerance is applied on either side of the basic profile.

18. A _____ profile is specified to control a line element of a surface.

19. Profile tolerance is a method of specifying control of deviation from the desired basic _____ along the surface of a feature.

20. Runout is a composite form and location control of permissible error in the desired part surface during a complete _____ of the part around a datum axis.

VIRTUAL CONDITION

Introduction

This chapter introduces you to a very important concept concerning the mating parts of an assembly. Today, interchangeability of parts is more critical than ever before. Interchangeability includes those manufacturing situations where subassemblies may be shipped from one country to another for assembly, and also includes the replacement of a single part within an assembly. The required assembly and interchangeability of parts and assemblies can only be ensured when parts are accepted at virtual condition or better. The virtual condition of a part is the condition that defines the boundary of acceptability. This condition is the boundary established by the collective effect of size and geometric tolerance.

Virtual condition is explained and illustrated for both internal and external features. The principles of the modifiers are involved here also. If these are not clear, review them before attempting to comprehend the concept of virtual condition. The virtual condition concept is especially important when developing gaging. Gages must be made to the virtual condition boundary and the resultant condition boundary as derived from the material condition specified to accept features, but not to accept those worse than virtual condition. Mating parts are dimensioned with consideration for virtual condition. It is through proper dimensioning and gaging with virtual condition that 100% interchangeability and proper function are achieved.

APPLICATION

Virtual condition is not a control, but a condition of a feature as a result of size, geometric tolerance, and the modifiers. Virtual condition is the boundary (locus) at which features are no longer acceptable. The locus may violate Rule One, which in part states, "the bound-

ary of perfect form at MMC." In certain applications, the combined effect of the actual mating size, geometric tolerance, and the modifiers will exceed the boundary of perfect form. The following examples will illustrate how the boundary may be exceeded.

EXTERNAL FEATURE

The first indication that a virtual condition may exist for a feature is the size tolerance: It must be a feature of size. The feature must also be controlled with a geometric tolerance. The next indication is whenever a center line or an axis is being controlled. Figure 7-1 illustrates these two conditions.

The virtual condition of an external feature is a constant value equal to its maximum material condition size plus its applicable tolerance of form or location. For this pin, the virtual condition is equal to 12.7 plus the geometric tolerance of 0.5. The virtual condition is 13.2 (outer boundary, locus).

The resultant condition of an external feature is a variable value equal to its actual mating envelope size minus its applicable tolerance of form or location. For the pin in Figure 7-1, the resultant condition is equal to 12.7 minus 0.5. The resultant condition is 12.2 (inner boundary, locus).

The outer boundary of this pin does exceed the boundary of perfect form according to Rule One. This is permitted because Rule One establishes the bound-

Figure 7-1

Figure 7-2

Figure 7-3

ary of perfect form for only those features that are not controlled with a geometric form tolerance. This pin then may occupy a diametral area up to, but not exceeding, 13.2 mm. This does not mean that the pin diameter may exceed the stated size and tolerance. If the pin does not meet the specified size requirements, it is not acceptable. It means that with the pin at the largest diameter (12.7) and the axis requiring the entire tolerance zone (0.5), the effective diameter, outer boundary is 13.2 mm. The actual mating size and geometric tolerance are within specification.

The designer must always consider virtual condition when dimensioning and tolerancing functional parts, because parts are inspected with functional gaging at virtual condition that represents the mating part. The functional gage for the pin in Figure 7-1 is illustrated in Figure 7-2.

This functional gage diameter must be 13.2 mm in order to check all acceptable pin shapes. This gage is not used to check feature diametral size. Figure 7-3 shows how pin forms are verified with a functional gage. When the pin is at the maximum diameter and at perfect form, it will fit the gage similar to that shown in Figure 7-3.

When the pin is manufactured at MMC (12.7 mm), the axis is permitted to lie within a diametral tolerance zone of 0.5 mm. This means the pin may also bow or wave to that extent. Figure 7-4 illustrates how this pin would be accepted by the functional gage.

The pin may also be produced to the other actual mating size limit (LMC). In this condition, the tolerance zone for the axis would increase to 0.9 mm. The pin may take any shape as long as the axis lies within the tolerance zone of 0.9. Figure 7-5 illustrates how

pin straightness could vary up to as much as 0.9 mm at the axis and still be acceptable.

PARALLELISM

The concept of virtual condition is not too difficult to follow with one feature. The application on a more realistic part will help to expand on this concept. The drawing in Figure 7-6 has two features that must mate

Figure 7-4

Figure 7-5

Figure 7-6

Figure 7-8

with another part, the other part being the functional gage.

The two pins in this part are subject to virtual condition. The lower pin, because it is the datum feature, must be simulated with a gage to establish the datum axis. In such cases, the datum feature applies at its virtual condition even though it is referenced in a feature control frame at MMC or LMC. The upper pin must also be considered with a virtual condition gage because it meets the requirements of virtual condition.

The gage cylinder for the lower pin would have to be adjustable to fit the produced condition of the pin. The gage cylinder for the upper pin will have a virtual condition diameter of 13.7 mm. The virtual condition is determined by adding the 0.5 mm tolerance in Figure 7-1 to the MMC size of 13.2 mm in Figure 7-5. The gage must be capable of accepting the entire length of the pins. Figure 7-7 illustrates the gage for the part in Figure 7-6.

Figure 7-7

INTERNAL FEATURE

Virtual condition for an internal feature must have the same conditions as an external feature. The feature must have size, a geometric tolerance and a modifier. The virtual condition is calculated opposite of an external feature. Virtual condition is the feature MMC size minus the geometric tolerance. Figure 7-8 illustrates a feature of size (hole), the geometric tolerance and a modifier.

The virtual condition of an internal feature is a constant value equal to its maximum material condition size minus its applicable tolerance of form or location. For this hole, the virtual condition is equal to 12.4 mm (MMC) minus 0.2 mm. The virtual condition is 12.2 mm (inner boundary, locus).

The resultant condition of an internal feature is a variable value equal to its actual mating size plus its applicable tolerance of form or location.

The virtual condition of this hole is smaller than the specified size limits of the hole. This is acceptable because the hole at any cross-sectional measurement is within the size limits and the axis may lie any place within the diametral tolerance zone of 0.2 mm at MMC. This hole, like the pin in earlier examples, is permitted to bow, taper, or wave. The shaft that passes through this hole may not be the same shape as the hole, therefore, some acceptable condition must be established—the virtual condition. Figure 7-9 illustrates how the hole may be produced.

Figure 7-9

Figure 7-10

The functional gage used to check this hole would have a 12.2-mm-diameter pin (fixed). The pin is not expandable as it would be if the tolerance was specified at RFS. The pin is positioned in the fixture so that it would also check the hole in relation to the datums. Figure 7-10 illustrates how this part might fit a gage for checking the hole.

This gage then will only accept parts that are 100% interchangeable if properly designed. The gage accepts holes of any configuration that are within the stated tolerances. This part will need two gages or a gage for the hole and an optical comparison for the profile in order to check and accept the part.

When the controlled feature is also a datum, as is the case with the hole in Figure 7-8, the datum feature applies at its virtual condition. The gage-pin axis is considered theoretically exact of the purpose of locating the part profile. When a gage is used to check a datum feature, the gage axis or center line becomes the simulated axis.

Summary

Virtual condition is a critical concept of GD&T from design through inspection. It is with proper application of GD&T and interpretation of the virtual condition concept that parts are 100% interchangeable. The concept of virtual condition may best be remembered as an extension of the modifier principles. Virtual condition is always determined to be the MMC size plus geometric tolerance for external features; in other words, the feature's largest effective actual mating size. For an internal feature the virtual condition is MMC minus the geometric tolerance. This would be the feature's smallest effective actual mating size.

Chapter 7 Evaluation

1. Virtual condition is a boundary of perfect form and orientation, generated by the _____ effects of the feature MMC size and applicable geometric tolerances.

2. To determine the virtual condition for an external feature, the feature's _____ tolerance must be added to the feature MMC size.

3. A feature at virtual condition may _____ the boundary of perfect form as required by Rule One.

4. Virtual condition is the condition in which _____ are made.

5. Virtual condition is the boundary at which features are no longer an _____ part.

6. Functional gaging at virtual condition represents the _____ part.

7. Virtual condition for an internal feature is calculated the _____ of that for an external feature.

8. Parts and features made at virtual condition or better are 100% _____.

9. According to Rule _____, the virtual condition of a feature of size is considered the simulated datum.

10. Virtual condition may also be considered as an extension of the _____ principles.

TOLERANCES OF LOCATION

Introduction

This chapter introduces the principles of tolerances of location. These tolerances or geometric controls are concentricity, symmetry, and position. These location controls are specified to control the relationships between features or between features and a datum feature. The relationships are toleranced at the axis, center line, or center plane. Positional tolerance then provides the permissible variation in the specified location of the feature or group of features in relation to another feature or datum. A tolerance of location is applied to at least two features of which one must be a feature of size (meaning one is a datum).

Because one of the features must be a feature of size, the modifier principles do apply. General Rule Two requires the designer to specify modifiers for all features, tolerances, and datums of size. The advantages of the modifiers can be used to their greatest extent with tolerances of location involving part interchangeability and functionality of mating parts. GD&T's best advantages are best realized when position and modifiers are specified.

Concentricity

SYMBOL

Figure 8-1

DEFINITION

Concentricity is the condition where the median points of all diametrically opposed elements of a figure or revolution (or correspondingly located elements of two or more radially disposed features) are congruent with the axis (or center point) of a datum feature.

TOLERANCE

Concentricity tolerance is always implied and specified as RFS, regardless of feature size. The tolerance is a diametral zone in which the axis of the controlled feature must lie. This zone must coincide with the axis (center point) of the datum feature(s). Concentricity is a very restrictive geometric control. A specified tolerance controls the amount of eccentricity error, parallelism of axis, out-of-straightness of axis, out-of-circularity, out-of-cylindricity, and any other possible errors in the feature axis. This tolerance controls all possible errors at the feature axis; therefore, it is difficult to verify and may be excessively expensive. The actual feature axes must lie within the specified tolerance zone.

APPLICATION

Concentricity is considered when critical axis-to-axis control is required for dynamically balanced features. This control is selectively specified, because features may be controlled with runout or position. Runout only controls a surface-to-axis at RFS. Position offers all the advantages of GD&T. Concentricity is an axis-to-axis control at RFS. Concentricity is normally specified for high-speed rotating parts, rotating mass, axis-to-axis precision, or any other feature critical to function. Figure 8-2 illustrates the proper application of concentricity.

TOLERANCE INTERPRETATION

Concentricity tolerance is always interpreted as RFS. The tolerance zone is diametral around and parallel to the datum axis. Verification is difficult. Concentricity is the determination of the controlled features axes in

Figure 8-2

Symmetry

SYMBOL

Figure 8-4

DEFINITION

Symmetry is the condition where a feature or part has the same profile on either side of the center plane (median plane) of a datum feature.

TOLERANCE

The tolerance for symmetry is always implied to be RFS. The tolerance is applied equally on either side of the controlled feature center line. The implied modifier restricts the tolerance to the specified amount only.

APPLICATION

Symmetry is specified for features to be located symmetrically with respect to the median plane of a datum feature. It may also be specified for a feature in a common plane with a datum plane. This symbol is specified extensively throughout industry. Figure 8-5 illustrates an application of symmetry.

relationship to the datum axis. A radial differential measurement is the most accurate method of determining the actual controlled feature axis. These measurements must be taken opposite of each other. The tolerance zone for Figure 8-2 is illustrated in Figure 8-3.

For concentricity, the feature surfaces must be measured diametrically opposed to each other to determine the midpoints relative to the axis of the datum.

For coaxial features located with a positional tolerance, the location of the axis of the features actual mating envelope is controlled relative to the axis of the datum feature.

Figure 8-3

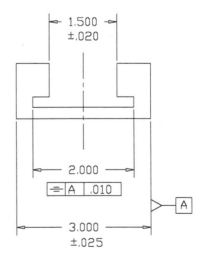

Figure 8-5

TOLERANCE INTERPRETATION

The specified tolerance establishes a tolerance zone that is the specified tolerance in width, with half of the tolerance on either side of the datum feature center line. This width zone allows the feature to vary from side-to-side or angularly within the tolerance zone. The tolerance zone size is not permitted to vary with feature size. Figure 8-6 illustrates how the interior of the part could vary in relation to the exterior, which is the datum feature. The controlled feature is permitted a maximum of .005 inch shift to the side in either direction.

Position

INTRODUCTION

Position is one of the most effective and used controls in GD&T. This control provides the designer with the ability to specify clearly all design requirements and intentions. Through clearer specification, higher production yields are possible, interchangeability is ensured, and definable quality requirements are attained. The coordinate method does not provide the same advantages. The coordinate system is not replaced with GD&T, but is enhanced with it. The two methods of dimensioning and tolerancing may be specified simultaneously on a drawing. As we stated, the coordinate method doesn't provide the same tolerance advantages; this is illustrated in Figures 8-7 through 8-11.

Position is always specified for features of size. Datums must be specified. Positional tolerance en-

ables the designer to specify geometric controls that utilize the majority of the advantages for specifying GD&T. Position may be specified to control feature locations, relationships, coaxiality, concentricity, and symmetry of cylindrical and noncylindrical features of size. Because position is specified for "size" features, the modifier principles must be considered. These principles do provide many of the advantages for specifying GD&T. The datum reference frame becomes an important concept when position is specified in relationship to datums. Position is always specified in conjunction with basic dimensions from specified datums or between interrelated features. Basic dimensions establish true position.

Positional tolerance may be explained either in terms of the internal surface of a hole, slot, etc., or in terms of the axis, center plane, or center line. This book, like most documents on the subject, explains position in terms of feature centers. The specified positional tolerance defines a zone within which the center of the feature of size is permitted to vary from true position (theoretically exact) in relation to another feature or datum. True position is established by basic dimensions from specified datum features and between interrelated features. True position is an axis, center line, or center plane of a feature as defined by basic dimensions. The tolerance specified in the feature control frame is either a diameter or a width located equally around true position.

The tolerance zone may be modified with the geometric modifiers. Application of the modifiers allows the specified tolerance to increase an amount equal to the actual mating size departure from MMC (maximum material condition) or LMC (least material condition).

Position Theory

INTRODUCTION

Design requirements for assemblies with interfacing mating parts usually relate one feature to another. These features, holes and pins, or holes for floating fasteners, relate to each other in 360° of location to each other. The coordinate method of tolerancing *does not* permit 360° flexibility. The coordinate sys-

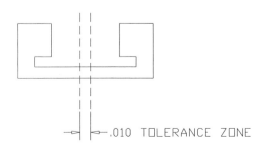

.010 TOLERANCE ZONE

Figure 8-6

Figure 8-7

NOTE: .060 TOLERANCE ZONE ONLY

Figure 8-8

tem makes *no* provisions for feature "size" in relation to feature "location." GD&T takes into consideration both "size" and "location" when determining feature or part acceptability.

APPLICATION

The following is an explanation of the reasoning behind position versus the coordinate system. For discussion and illustration purposes consider the requirement for a .530 plus .008, minus .002 diameter hole in a part. The hole is to be located within plus or minus .030 to the center of the part. This translates to a .060 square tolerance zone around true position for the actual axis of the hole to lie within. Figure 8-7 illustrates the drawing requirement.

AXIS LOCATION

The axis of the hole must lie within the square zone of .060 to be an acceptable part. To further expand upon this, assume that five parts were inspected, and the axis of each hole is illustrated in relation to the .060 tolerance zone in Figure 8-8. The illustration shows that two parts are acceptable and three are out-of-tolerance.

DIAGONAL MEASUREMENT

The concept behind positional tolerancing is to use a circular tolerance zone rather than a square one.

The circular zone allows for 360° of axis or feature movement. To achieve the circle, the diagonal measurement from center to a corner becomes the radius for a circular zone. In this illustration the diagonal measurement is .042, so if part number two is acceptable, then part three should also be acceptable because it is closer to true position. By specifying a circular tolerance zone, four of the five parts are acceptable. Figure 8-9 illustrates the two tolerance zones.

CIRCULAR TOLERANCE

In Figure 8-9 the illustrated approach to tolerancing appears to be more logical, but another consideration must be made. The designer intended for the holes to vary in the *x* and *y* directions, not diagonally; therefore, the circular tolerance zone must not exceed the original intention of .030. The tolerance zone must be an inscribed circle .060 in diameter. Figure 8-10 illustrates the proper comparison between the coordinate and geometric tolerance zones. This illustration appears to have reduced the allowable variation in feature location; in fact, it *has*, for a feature of minimum size. When the modifier principles are considered, the tolerance is not lost, but gained.

Figure 8-9

Figure 8-11

BONUS TOLERANCE

One of the advantages of GD&T is a greater tolerance based on actual mating size. In this example, assume parts four and five were produced at the largest permissible size, .538. The preceding .060 tolerance zone has increased to .070. This practice is possible with the application of the modifier principles. The ASME Y14.5M-1994 standard states in part that "the tolerance is limited to the specified value when the feature is at MMC. Where the actual mating size departs from MMC, an increase in tolerance is allowed equal to the amount of departure." Figure 8-11 illustrates how the bonus tolerance zone accepts parts four and five when they are made larger than the

specified MMC size. All of the parts are acceptable if their axes remain in this .070 tolerance zone for the thickness of the part. This tolerance zone can be considered to be a tube (see Figure 8-14). The axis of the controlled feature must lie in the tube in any shape or manner as long as it is in there. For noncylindrical features, the zone would be like a box or rectangle.

Position

SYMBOL

Figure 8-12

DEFINITION

Position is the condition where a feature or group of features is located (positioned) in relation to another feature or datum feature.

Figure 8-10

TOLERANCE

Location tolerance zones are either cylindrical or noncylindrical; this is determined in the feature control frame. The tolerance zone is cylindrical if the diameter symbol precedes the specified tolerance. The absence of the symbol indicates a width zone. The controlled feature(s) axis or center line must lie within the allowable tolerance zone, which is equally distributed around true position. True position is a theoretically exact location determined with basic dimensions. The specified tolerance zone may increase in size based on the actual mating size. Tolerance zone increase is permitted with the specification of modifiers.

APPLICATION

Position may be specified to control nearly all features of a part. Position should be specified whenever the design requirements permit. This control provides an opportunity to utilize many of the advantages of GD&T. Figure 8-13 illustrates a simplified application of position.

TOLERANCE INTERPRETATION

The tolerance in this example is .030 in diameter when the feature (hole) is at MMC, .528. The tolerance zone is cylindrical because the designer specified the diameter symbol preceding the tolerance. Figure 8-14 illustrates how the tolerance zone is determined. The tolerance zone in this position example is permitted an

Figure 8-14

increase in size by .010. This is available based on the feature plus/minus size tolerance. The tolerance zone diameter is determined by actual mating size. Before the feature is produced, only the minimum and maximum tolerance zone sizes can be determined. If this example contained a group of features, each of them may have a different size tolerance zone based on the actual mating size.

Positional tolerance specified at MMC may be explained in terms of the surface of a hole or the axis of a hole.

a) *In Terms of the Surface of a Hole*—While maintaining the specified size limits of the hole, no element of the surface is permitted to be inside the theoretical boundary located at true position.

b) *In Terms of the Axis of a Hole*—Where a hole is at MMC (minimum diameter), its axis must lie within a cylindrical tolerance zone whose axis is located at true position. This zone diameter is equal to the positional tolerance specified for the hole.

Note: In certain cases of extreme from deviation (within limits of size) or orientation deviation of the hole, the tolerance in terms of the axis may not be exactly equivalent to the tolerance in terms of the surface. In such cases, the surface interpretation shall take precedence.

Position tolerance in this book is explained in terms of the axis.

Figure 8-13

Position of Multiple Cylindrical Features

INTRODUCTION

The advantages of GD&T and position can best be explained when two or more features are positioned. When a group of features are to be controlled with position, a tolerance for location is specified and appropriately modified. The feature is then related to something, which may be an edge, surface, or other feature(s); these are the datum reference features. Specification in this manner ensures a clearer intent of design. The datum reference frame is utilized for the relationships as required for part function. The rules are applied as required for the specified controls and controlled features. The geometric control—position—brings all of these concepts together better than any of the other controls. An illustration of how multiple features are specified and controlled with position is shown in Figure 8-15.

FEATURE LOCATION

The effects of feature size, MMC, and virtual condition can best be illustrated when the explanation begins with the features perfectly located at MMC. Figure 8-16 illustrates these holes with the gage inserted. Note that the gage pins are perfectly centered in the holes when they are in this condition. There is clearance

GAGE PINS
AT .498

Figure 8-16

between the pins and holes because the gage is at virtual condition. For this multiple-feature example the gage pins are .498 in diameter. When all features are located perfectly as they are here, there will be an equal distance between each gage pin and the hole. This gage is simulating the fit or assembly of a mating part.

OPPOSITE OFFSET AT MMC

The gage will accept all acceptably located features regardless of their shape, location, and size (must be verified separately). Each feature may be off true position in a different direction and be of different sizes. Figure 8-17 illustrates a possible hole arrangement. These features are at their maximum tolerance offset at MMC. The axis of each feature is .015 of an inch from true position.

The gage will accept this part when the features are offset in various directions. Figure 8-18 illustrates the part and gage. Note there is line contact between the gage and part.

Figure 8-15

NOTE: HOLES ARE ALL .528 DIA.
AND OFFSET.

Figure 8-17

NOTE: HOLES ARE ALL .528 DIA.
THE GAGE PINS ARE .498

Figure 8-18

GAGE PINS ARE .498 DIA.

Figure 8-20

OPPOSITE OFFSET AT LMC

The three features are more likely to be produced at some size larger than MMC. Normally, holes and other features are produced at or near their nominal size or larger. In Figure 8-19 the outside two holes are at their maximum opposite offset at LMC, and the center one is at the maximum vertical offset at MMC. When the holes are produced at LMC, as they are in this illustration, the advantage is "maximum tolerances" or bonus tolerance. Here an internal feature is produced at its largest specified size. In this condition the additional feature size increase beyond MMC may be added to the specified location tolerance; therefore, for the two outside holes, the location tolerance is .040 of an inch, a bonus of .010 over MMC.

The gage will accept the part in Figure 8-19. All features are within specified size and location tolerances. Figure 8-20 illustrates the part and gage. Note

that there is a line contact between the gage and features. The examples illustrated in Figure 8-18 and 8-20 or any other conceivable combination illustrate part acceptance when manufactured to the specifications. The parts are acceptable if they were thin sheet-metal parts or thick castings. The feature axis must lie within the "tube" of tolerance for the thickness of the part.

Composite Positional Tolerancing

INTRODUCTION

Composite positional tolerancing is usually specified when a feature-to-feature relationship is more critical than the pattern relationship to the part edges. Composite tolerancing is the combining of two or more feature control frames for the purpose of locating a pattern of features to a three-plane datum reference and then controlling the feature-to-feature relationship within the pattern. Composite tolerancing allows the designer the flexibility of specifying the required precision for pattern location and features within the pattern. The modifiers are usually specified to gain the advantages provided by them.

Where design requirements permit composite positional tolerancing, a feature-relating tolerance zone framework (FRTZF) to be located and oriented within the limits imposed by the pattern-locating tolerance zone framework (PLTZF) is used. (The acronyms are pronounced "fritz" and "plahtz.")

Figure 8-19

Figure 8-21

Figure 8-22

COMPOSITE CONTROL

The combined feature control frame shown in Figure 8-21 provides composite positional control. The first line (PLTZF) controls the location of a pattern by basic dimensions in relationship to the specified datum reference frame. This PLTZF provides pattern control for shift or rotation in the part. The second line (FRTZF) specifies the positional tolerance to control feature-to-feature relationship within the pattern. This tolerance allows individual features to vary from true position within the PLTZF. The tolerance establishes individual tolerance zones for each feature in the pattern. The tolerance controls feature-to-feature relationship and feature attitude or perpendicularity as it passes through the part. Usually, the same datums and datum precedence are *not* specified for the individual feature control. Where datum references are specified, one or more of the datums specified in the upper segment of the frame are repeated, as applicable, and in the same order of precedence, to govern the orientation of the FRTZF.

PATTERN CONTROL

PLTZF is specified to provide an orientation requirement. The control only specifies where the pattern is to be located in relationship to the datum references. The PLTZF is specified with basic dimensions from specified datums, as illustrated in Figure 8-22. The features within the pattern are also spaced with basic dimensions. These dimensions provide the *true position* for the pattern and individual features. The specified tolerance of 0.030 applies at the true position for each individual feature in the pattern. This tolerance zone can be thought of as a dart board or target. This is the area that must be hit for the feature to be acceptable. The hit is as good in one place as it is in another, including true position.

The designer permits the use of a 0.030 area; use it! If the hit or feature had to be at true position, parts would be excessively expensive. When feature accuracy increases, so does the cost of manufacturing to achieve that accuracy. This type of tolerancing is not to promote ill fitting or poor appearing parts and assemblies. If the designer has calculated the tolerance, use what is allowed.

FEATURE CONTROL

The FRTZF is established by the second line of the combined feature control frame. The tolerance specified here controls the feature-to-datum and feature-to-feature relationship (Figure 8-23). The specified tolerance applies individually to each feature in the pattern and they *must* be at the basic dimensions from each other. This FRTZF tolerance zone must fall within, or at least make a line-to-line contact with, the PLTZF (Figure 8-24). The axis of each individual feature must lie within this tolerance zone. (**Note:** Example "A" in Figure 8-24 is not questionable. Ex-

Figure 8-23

Figure 8-24

Figure 8-26

Figure 8-27

ample "B" is difficult to verify without a gage and the axis must lie at the point of line-to-line contact.)

AXIS VERIFICATION

The axis of each individual feature must lie within *both* tolerance zones simultaneously of the feature to be acceptable. Figure 8-25 illustrates how the tolerance zones and feature axes may appear. Each line of the feature control frame would require separate consideration. By specifying only datum "D," the tolerance zone is perpendicular to that surface, controlling both location and perpendicularity of the FRTZF.

Two Single-Segment Feature Control Frames

INTRODUCTION

Single-segment feature control frames are used when it is desirable to require basic dimensions along with the datum references. This method of positional

tolerancing is also used to position a pattern of features in relation to the three-plane datum concept.

CONTROL

The two single-segment feature control frames are illustrated in Figure 8-26.

The lower feature control frame repeats datums "D" and "E." Figure 8-27 shows that the tolerance cylinders of the FRTZF (as a group) are free to be displaced to the left or right as controlled by the basically located cylinders of tolerance of the PLTZF. These tolerance cylinders are required to be perpendicular to plane "D" and parallel to "E."

The actual axes of the holes in the pattern must lie within both the tolerance zones of the FRTZF and PLTZF, as in Figure 8-28.

Figure 8-25

Figure 8-28

Multiple Patterns Located by Basic Dimensions and Related to the Same Datums

INTRODUCTION

Many of the parts throughout the industrial world contain more than one pattern of features. These patterns may be related so that they must be verified together. If the patterns can be separate, they are so specified to reduce the part cost. It is not necessary for the designer to always add the note "SEP REQT." Separate requirements are also specified by different

datum references, different modifiers, and different datum order of precedence.

PATTERNS LOCATED RELATIVE TO DATUMS OF NO SIZE

When multiple patterns of features are located in relation to common datum features not subject to size variation, or to common datum features at RFS, they are considered to be a single pattern.

In Figure 8-29 two patterns of features located from common datums not subject to size variation are illustrated. Because both patterns are located from these datums with basic dimensions, the patterns are

Figure 8-29

Figure 8-30

PATTERNS LOCATED RELATIVE TO DATUMS OF SIZE

verified with one setup or gage. When patterns are specified in this manner, they are *not* allowed any shift or rotation independent of the other pattern.

When patterns of features are located from the same datums of size, verification again is performed as if the patterns were a single composite pattern. The feature control frames must contain the same datums in the same order of precedence and with the same modifiers. If the design does not require such an interrelationship between patterns, the designer may add the notation SEP REQT (separate requirement) under each applicable feature control frame. This notation allows each pattern to be verified independently. It also allows pattern shift/rotation independent of each other. Figure 8-30 illustrates an independent design.

The designer may change the contents of the feature control frames. The designer may designate different datum references, order of precedence, or modifiers, thereby relieving the composite pattern requirement. Patterns of features are independent whenever feature control frames are different or when the notation SEP REQT is added beneath the feature control frames.

Patterns Positioned from a Datum of Size

INTRODUCTION

Positional tolerances may be specified to control a pattern of features in relationship to another feature. This type of specification is used when pattern-to-feature relationship is more critical than pattern-to-edge. The single feature in this type of application becomes the locating feature for the pattern. Most frequently, the locating feature is a feature of size. Features of size must have a modifier specified of them in the feature control frame. These locating features are also controlled by Rule Four, the datum/virtual condition rule. The specification of modifiers in these designs provides the full advantages of GD&T. If RFS (regardless of feature size) is specified, the tolerances are more restrictive, and verification is more difficult.

REGARDLESS OF FEATURE SIZE

A locating feature (datum) for a related pattern is shown in Figure 8-31. In this example, the locating datum feature for the pattern is modified to RFS. This

Figure 8-31

is restrictive, but some designs do require the restriction based on the function of the final assembly.

To begin, the center hole must be located as specified by the basic dimensions and feature control frame. The hole is to be 1.500 in. plus .008 in. or minus .002 in. The hole is allowed a .020-in . diametral tolerance zone for location. In this example the positional tolerance zone for the hole is modified to MMC. This means that

the hole has a .020-in. tolerance when it is produced at 1.498 in. If the hole is produced at 1.508 in , the tolerance zone increases to .030 in. The hole was measured to be 1.502 in., so the positional tolerance is .024 in.

Next, the pattern of eight holes must be produced in relationship to the center hole, which has become datum feature "X." Datum feature "X" is modified to

Figure 8-32

Figure 8-33

RFS in the feature control frame for the eight holes. These eight holes are to be .375 in., plus .008 in. or minus .002 in. They each have a positional diametral tolerance of .005 in. at MMC. Figure 8-32 illustrates the possible tolerance zones for all of the features.

In this example, as presented with datum feature "X" at RFS, the pattern shift is restricted to zero. The shift is zero because regardless of the feature location or size, it becomes the dimension origin for the true position of the pattern of eight holes.

Pattern Tolerance When a pattern of features is located from an RFS datum, the pattern is *not* permitted any shift from the datum feature. Here the datum feature axis becomes the origin for the pattern-locating dimensions. Figure 8-33 illustrates the dimensioning for the pattern from datum feature "X." All of these dimensions must be basic.

These basic dimensions establish the true position locations for the pattern. When the basic true position of the pattern in relationship to datum feature "X" is established, the individual tolerance for each feature in the pattern allows variation from true position. Each feature has an individual tolerance based on the actual mating size. This tolerance may vary from .005 to .015 in. because of the MMC modifier following the .005-in. positional tolerance. Each feature in the pattern is permitted to shift or vary as the three features did in

Figure 8-23. Verification of this part would require the establishment of the datum axis. This axis must meet the drawing specifications. After verifying an acceptable datum axis, proper pattern location must be measured. If the pattern is properly located, each feature within that pattern must be verified for proper location and orientation (perpendicularity to datum surface "A"). Verification may be accomplished, depending on required accuracy, with a coordinate measuring machine, paper gaging, or one hard gage that has an adjustable pin to fit the datum feature. The pin must be adjustable because of RFS.

MAXIMUM MATERIAL CONDITION

When any of the common datums in multiple patterns of features are specified at MMC, there is an option whether the patterns are to be considered as a single pattern or as having separate requirements. If no note is added under the feature control frames, the patterns are treated as one. When the patterns are to be treated individually, a notation SEP REQT (separate requirement) is placed beneath each feature control frame.

The MMC modifier specified for a datum feature of size permits the pattern to vary depending on datum size. When the datum feature is at MMC the pattern variation is restricted, but as the datum feature increases

Figure 8-34

in size, the amount of pattern orientation shift/rotation also increases. Tolerancing patterns in this manner is more common than RFS. Figure 8-34 specifies positional tolerances at MMC for the same part discussed with RFS.

The center hole or datum feature "X" must be located with basic dimensions. The hole must meet the specified size and location requirements. The hole size may vary from 1.498 to 1.508 in. with a positional tolerance that is .020 in. at MMC to .030 in. at LMC. For this example, the hole measures 1.502 in., which is .004 larger than MMC, plus the .020 positional tolerance, equals a .024 positional tolerance.

The pattern of eight holes must be produced in relationship to the axis of datum feature "X." The feature control frame for the eight holes contains datum reference letter "X" with a MMC modifier. This modifier permits some pattern shift/rotation, an amount equal to the feature size departure from MMC. This amount of departure is also the amount the pattern may vary from true position. Figure 8-35 illustrates the available pattern shift/rotation tolerance. The illustration contains a gage pin at virtual condition for datum feature "X." This datum feature is controlled by Rule Four, datum/virtual condition. The gage pin measures 1.478 in.

Pattern Tolerance The pattern shift/rotation is achieved by the amount of clearance there is between the hole and gage pin. The part may be shifted/rotated all around the gage pin in relationship to the other datum features (part edges).

The eight holes in the pattern establish a true position in relationship to the gage axis. When the pattern orientation is established from the simulated datum axis, each of the eight features has an individual positional diametral tolerance of .005 in. at MMC. This

Figure 8-35

Figure 8-36

tolerance permits each feature variation based on actual mating size. Figure 8-36 illustrates individual feature tolerance in relationship to the basic pattern orientation.

The MMC modifier permits the pattern some shift/rotation in relationship to the datum feature. Then each individual feature in the pattern is permitted variation for true position. The amount of variation is dependent on the actual mating size. For example, if the datum feature is produced at the virtual condition size, there is no pattern shift/rotation permitted. Likewise, if any or all of the eight features were produced at virtual condition, they would have to be located at their true position.

Parts that have patterns of features related to another datum feature(s) may be verified with several methods. Setup time and accuracy required may determine the method of verification. These parts may be verified with a coordinate measuring machine, paper gaging or hard gaging. Each of these methods provides advantages and disadvantages. The datum feature must be verified first with any method. Separate gaging is required to determine acceptable location. Then from the actual datum feature axis the pattern location is determined. Another gage is required to verify pattern relationship to the datum feature axis.

Zero Tolerancing

INTRODUCTION

Zero positional tolerancing at MMC may be specified where design requirements permit. Additional tolerance then would be allowed when a tolerance is specified in the feature control frame. Zero tolerancing *does not* indicate no tolerance or less tolerance. This method of tolerancing may allow more tolerance and tooling options. The additional tolerance is achieved by the designer adjusting the feature size. The virtual condition for the features function is specified as one limit of size with a plus or minus size tolerance depending on an internal or external feature. When the feature is produced at MMC, it must be at true position with zero tolerance.

APPLICATION

To illustrate the advantages of zero tolerancing, the drawing in Figure 8-16 will be used. In that figure, the feature size and positional tolerance is specified, as shown in Figure 8-37.

That specification is a practical specification for a 0.500-in. fastener. The same hole limits of size and positional tolerance can be achieved with a zero tolerance specification. If zero tolerancing were specified, it would appear as illustrated in Figure 8-38.

COMPARISON

This method of tolerancing may at first appear as if there is no positional tolerance. The feature at MMC does not have any positional tolerance. However, as the feature departs from MMC, the tolerance zone increases in size from .000 to .040 in. With zero tolerancing, the feature size may vary from .498 to .538 in. In Figure 8-14 the features could vary in size from .528 to .538 in. with a .040-in. tolerance maximum. Zero tolerancing allows the features to be produced at their

Figure 8-37   *Figure 8-38*

FEATURE SIZE	POSITIONAL TOLERANCE		DRILL SIZE
.528	.030	= MMC	17/32
.531	.033		
.538	.040	= LMC	
.498	.000	= MMC/V.C	
.500	.002		1/2
.516	.018		33/64
.531	.033		17/32
.538	.040	= LMC	

Figure 8-39

minimum functional size or virtual condition and allows for a larger range of tool sizes.

VIRTUAL CONDITION

Zero positional tolerance is based on feature virtual condition. The designer must calculate the virtual condition for parts that assemble, so they may just as well specify that size and provide maximum tolerances for manufacturing. Figure 8-39 is a table of available drill or punch sizes to produce the features required by the two drawing specifications.

The comparison table in Figure 8-39 shows the flexibility that is available with zero tolerancing. Also, the lower size limit for an internal feature is both MMC and virtual condition.

DEFINITION

The projected tolerance zone specifies a required tolerance to prevent interference of mating parts when fixed fasteners are used. The required specified tolerance is projected above the surface that contains the fixed fastener.

TOLERANCE

Projected tolerance zones are usually cylindrical, but may be noncylindrical if the design permits. The tolerance is specified in a combined feature control frame. The upper part of the combined frame specifies the positional tolerance in relationship to the required datums. Then the lower part specifies the height of the tolerance zone above the part that is to contain the fixed fastener.

APPLICATION

A projected tolerance may be specified in an application where one part of an assembly contains the pins or studs to locate another part, cap, or cover to the first. Figure 8-41 illustrates such an application. The designer here has specified a projected tolerance of .500 in. above the part containing the fixed fastener. This height is required to ensure assembly of the mating part. The tolerance zone is a .015-in. cylinder 0.500 in. high at MMC.

TOLERANCE INTERPRETATION

The tolerance zone in this example is .015 in. at MMC above the part. This zone is specified to accommodate any error in the drilling or tapping process. The

Projected Tolerance Zone

SYMBOL

Figure 8-40

Figure 8-41

Figure 8-42

zone permits some misalignment of the two parts or some error in producing the holes. The illustration in Figure 8-42 shows the two holes produced at slight angles to each other. The axis of the hole in the mating part must be within the specified projected tolerance zone for the thickness of the part (.500 in.). When both parts are produced within their tolerances, the parts will assemble.

Noncylindrical Features

INTRODUCTION

Positional tolerancing may also be specified for noncylindrical features. The feature control frame is an indication that the feature may be noncylindrical; the diameter symbol will not be specified preceding the geometric tolerance. When the diameter symbol is not specified, the tolerance zone is then a *width* equally divided on either side of the feature center line.

The modifier principles apply to noncylindrical features with the same advantages as cylindrical features. Where modifiers are applicable, the tolerance zones are permitted to increase. The modifiers may also be specified for datum features of size. In such applications Rule Four, the datum/virtual condition rule, applies.

Noncylindrical features can be evaluated in the following three ways:

a) *In Terms of the Surfaces of the Feature.* No element of the feature's side surface shall be inside the theoretical boundary defined by two parallel planes (feature's specified minimum width) equal distance from the true position.

b) *In Terms of the Center Plane of the Feature.* The feature's center plane must be within the tolerance zone defined by two parallel planes (feature's tolerance) equal distance from true position. This tolerance zone is what controls the altitude (angle) that the feature must maintain.

c) *In Terms of the Boundary for an Elongated Feature.* No element of the feature's surface shall be inside the theoretical boundary of identical shape located at true position. The boundary size is equal to the MMC size of the elongated feature minus its positional tolerance. To require this concept, the term **BOUNDARY** is placed beneath the feature control frame.

Note: This boundary concept can also be applied to other irregularly shaped features such as D-shaped holes, where the center is not conveniently identifiable.

APPLICATION

Positional tolerance is applied to noncylindrical features in the same way it is applied to cylindrical features. Positional tolerance and datum references must be specified. The diameter symbol is not specified. The specified feature size and geometric control are usually attached to the controlled feature with extension lines. Figure 8-43 illustrates the application of a noncylindrical position control.

TOLERANCE

The specified positional tolerance of .005 in. at MMC is a *total width* tolerance for each spline of the feature when the spline is at .313 in. Because these splines are external features, the smaller they are produced, the larger the positional tolerance. In the example illustrated in Figure 8-44, the .005 in. tolerance is applied equally on either side of the spline center line. The center line may be off to one side, bowed at an angle, or any other possible form. It must lie within the tolerance zone for the feature

Figure 8-43

Figure 8-44

Figure 8-45

to be acceptable. This tolerance controls both the location and the form of each spline.

Bidirectional Tolerancing

INTRODUCTION

Bidirectional tolerancing is another method of positioning round holes. This method of tolerancing is applied when holes may be allowed to vary more in one direction than in another. Holes are located from either rectangular or polar coordinates. The principles apply to either method of coordinate dimensioning. The feature control frame does not contain a diameter symbol before the positional tolerance. The tolerance zone is rectangular and established by two different feature control frames. Bidirectional tolerancing is not limited to holes.

APPLICATION

Bidirectional tolerancing is applied in the same manner as other geometric controls. The feature control frames are attached to the controlled features in a drawing view where the feature relationships can be seen. One feature control frame provides the positional tolerance in one direction and the other one provides control in the opposite direction. Each of the tolerance values represents a distance between two parallel planes or arcs that is the specified tolerance apart. Figure 8-45 illustrates a rectangular coordinate method or bidirectional tolerancing.

TOLERANCE

The tolerance zones in each direction for bidirectional tolerancing are independent of each other. Combined, they form a rectangular zone. The zones are subject to change when MMC or LMC is specified in the feature control frame. The axes of the holes must lie within the allowable tolerance zones. The axis must remain within the zone for the thickness of the part. If a part is thick enough to where hole angularity or slant through the part is a concern, the designer may specify

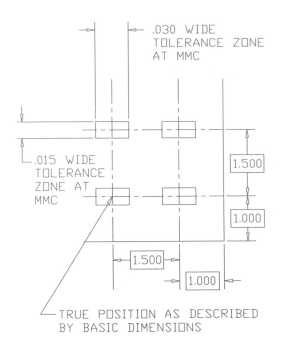

.030 WIDE
TOLERANCE ZONE
AT MMC

.015 WIDE
TOLERANCE
ZONE AT
MMC

1.500

1.000

1.500

1.000

TRUE POSITION AS DESCRIBED
BY BASIC DIMENSIONS

Figure 8-46

a perpendicularity or similar control to prevent excessive angularity. The tolerance zone for Figure 8-45 is illustrated in Figure 8-46.

Coaxial Features

INTRODUCTION

Coaxiality control is where the designer is controlling the axis-to-axis relationship of two or more cylindrical feature surfaces simultaneously. Coaxial features share the same principles as any other mating parts, therefore, datums and modifiers apply. Along with them are the rules governing features of size. Depending on the method of feature control, the advantages of GD&T can be realized when controlling coaxial features. The designer may specify one of three geometric controls. Depending on design requirements, coaxial features may be controlled with position, run-out or concentricity. Their application should be considered in this order.

POSITIONAL CONTROL

Coaxiality is most commonly controlled with position. Position provides the required interchangeability, potentially more tolerance, and permits the use of functional gaging. Whenever possible, the MMC modifier should be specified to achieve full advantage of tolerances. Positional control may be specified with the same positional tolerance and datums for both features, different positional tolerances, and the same datum feature references or different tolerances.

TOLERANCES AND DATUM REFERENCES

When the same positional tolerances and datum references are specified, a single feature control frame is used to control the features. Figure 8-47 illustrates this application. (**Note:** Same tolerances and datums.) When features are controlled in this manner, each of the controlled features is produced in accordance with the one feature control frame.

The specified control in Figure 8-47 requires that both the hole and the countersink axes lie within the same tolerance zone. The illustration in Figure 8-48 shows how the features may lie within the tolerance zone.

The design may require or allow for different tolerances for the counterbore and hole, but uses the same datum reference frame. This application requires two feature control frames: one for the hole pattern and the

Figure 8-47

Figure 8-48

Figure 8-50

other for the countersink pattern. Figure 8-49 illustrates this requirement. (**Note:** Different tolerances, same datums.) The two tolerance zones are coaxially located at true position as described by basic dimensions. The feature control frames follow the size specification of the controlled features.

The control specified in Figure 8-49 requires that the hole pattern lie within one tolerance zone and the countersinks lie within a different coaxial zone. These features, as specified, also constitute a composite pattern. If the design would allow separate requirements, the designer would have to add the note SEP REQT beneath each feature control frame. Figure 8-50 illustrates the tolerance zones as they were specified.

Designs that require the control of the countersink to the hole will have different tolerances and datums specified to control each group of features; therefore, each group of features will have a feature control frame specified. In addition, a note is required under the datum feature symbol for the hole and under the

feature control frame for the countersink, indicating the number of places each applies individually. Figure 8-51 illustrates this method of application. (**Note:** Different tolerances and datums.)

The method of controlling coaxial features with different tolerances and datums as just described is similar to specifying control for any two individual features. This specification requires the verification of the hole pattern without regard for the countersinks. When the hole pattern is verified acceptable, then each countersink-to-hole must be verified individually.

This method has two tolerance zones where one is positioned in relation to the axis of each individual hole. In Figure 8-52 the two tolerance zones are illustrated. The tolerance zone for the countersink is dependent on hole location. Each individual hole axis controls the location of each countersink.

Figure 8-49

Figure 8-51

Ø.003 AT MMC

Ø.005 AT MMC

NOTE: AXIS OF C'SINK IS DEPENDENT
UPON AXIS OF HOLE.

Figure 8-52

Control of coaxial features with position provides the maximum benefits of GD&T. All of the advantages of positional control, including zero tolerancing, may be specified when controlling coaxial features.

RUNOUT CONTROL

Runout control of coaxial features is an axis-to-surface control. Runout is normally specified where a combination of surfaces are cylindrical, conical, or spherical relative to a common datum axis. Runout is more restrictive than position because it is always implied as RFS. Runout controls were presented in detail in Chapter 6.

CONCENTRICITY CONTROL

Concentricity controls of coaxiality is an axis-to-axis control of all cross-sectional elements of a surface in relation to a common datum axis. Concentricity tolerance specifies a cylindrical tolerance of a specified diameter at RFS. Concentricity control requires the establishment and verification of all axes irrespective of surface conditions. Concentricity is usually specified as a last resort because it is difficult to verify. Concentricity was presented at length at the beginning of this chapter.

Summary

Location tolerancing is the most advantageous part of GD&T. When location tolerances are properly spec-

ified, the rules, modifier principles, datum references, and feature control frames can be applied to their greatest advantages. It is through proper application and interpretation that parts are 100% interchangeable, production yields increase, tolerances are maximum, functional gaging techniques can be implemented, production costs are reduced, among many more advantages.

Location tolerances are the specified amount of variation from true position or another feature. True position is established by basic dimensions from specified datum features. The specified tolerance provides a cylindrical or noncylindrical zone for the depth or length of the controlled feature. This zone is permitted to increase in size as the feature size varies from MMC. The actual feature axis or center line must lie within the zone. The axis or center line may take any form as long as it remains within the zone.

Position is the most frequently applied location tolerance. Concentricity should seldom be specified or it should be limited to features requiring dynamic balance. Position may be specified for nearly all features that require mating part assembly or interchangeability. This chapter should be studied to make certain the concepts are understood. Concentration on how the various concepts of GD&T are interrelated is important to the reader.

Chapter 8 Evaluation

1. The geometric tolerances of location are _____ and _____.

2. Tolerances of location specify the amount of _____ permitted between features and datum features or between features.

3. Concentricity is always specified or implied _____.

4. The tolerance zone for position is either _____ or _____ as determined from the feature control frame.

5. True position is specified by _____ dimensions.

6. Datum feature references must be _____ on drawings.

7. For an internal feature, position tolerance at MMC _____ as the feature approaches LMC.

8. When a composite features control frame is specified to locate a pattern of feature in relationship to a datum reference frame, the first line of the feature control frame controls pattern _____.

9. Concentricity tolerance is always specified in relationship to another datum _____.

10. Posiiton tolerance zones extend to the _____ or _____ of the controlled feature.

11. Datum features of size are considered at their _____ size when verifying patterns located from them.

12. Positional tolerances specified for features at _____ lend themselves to functional gaging verification.

13. Position tolerances specified at _____ provide the greatest advantages of GD&T.

14. Positional tolerances are specified for features of ____ _____.

15. Location tolerancing is a _____ method of controlling part features.

CHAPTER NINE

PRACTICAL APPLICATIONS

Introduction

This chapter is intended to provide the reader an opportunity to apply various GD&T concepts. The examples presented are based on ASME Y14.5-1994 practices and should be answered based on that standard. The material consists of a few form and orientation examples, but is primarily centered around positional tolerancing. Position is the most involved of the controls and is probably specified most frequently.

These review problems should be completed after reading the entire text; that way all of the GD&T concepts can be applied. This review may be used as a refresher or evaluation of existing GD&T knowledge.

SPECIFIED ON DRAWING

ACTUAL PART

Figure 9-1

Form and Orientation Tolerances

Referring to the drawing in Figure 9-1, answer questions 1–5.

1. When the feature control frame is specified as $\boxed{-\ |.005}$, what is the maximum for D _____ and E _____ ?

2. When the feature control frame is specified as $\boxed{\diagup\ |.003}$, what is the maximum for D _____ and E _____ ?

3. Which modifier is implied for these feature control frames? _____

4. Does one of the GD&T rules control the modifier application in these feature control frames? _____ If so, which one? _____

5. What is the maximum material condition size of this part? _____

Referring to the drawing in Figure 9-2, answer questions 6–10.

6. When the feature control frame is specified as $\boxed{0\ |.005}$, what is the maximum bow, bend, curve, wave, etc., permitted? _____

7. When the feature control frame is specified as $\boxed{\diagup\ |.002}$, what is the maximum bow, bend, curve, wave, etc., permitted? _____

8. If a geometric feature control frame was not specified for this part, is there any control for form? _____ If so, what? _____

9. What is the least material condition size of this part?

_____

SPECIFIED ON DRAWING

ACTUAL PART

TOLERANCE ZONE

Figure 9-2

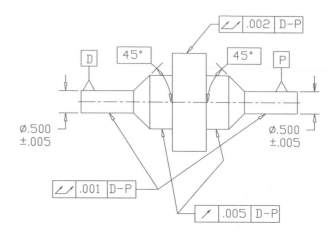

Figure 9-4

10. Could the designer specify MMC in these feature control frames? _____ If so, which rule is involved? _____

Referring to the drawing in Figure 9-3, answer questions 11–14.

11. The bottom surface of this part is identified as _____ "A".

12. Is the tolerance zone for the .252 hole cylindrical or noncylindrical? _____ Why? _____ _____

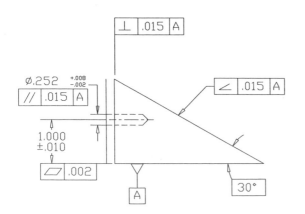

Figure 9-3

13. Is the tolerance zone for the 30° angle a width or an angularly shaped zone along the face of the angle? _____

14. Can any of the geometric controls in this drawing be modified to MMC? _____ If so, which? _____

Referring to the drawing in Figure 9-4, answer questions 15–19.

15. Runout is a _____-to-axis control.

16. Which of the two runout controls specified is more stringent? _____

17. Runout tolerances are always RFS and normally implied to be a _____ reading.

18. All surface elements are controlled simultaneously with _____ runout.

19. The controlled features must meet the requirements of Rule _____.

Referring to the drawing in Figure 9-5, answer questions 20–25.

20. Concentricity is an _____-to-axis control of two or more features.

Figure 9-5

Figure 9-6

21. When concentricity is specified, surface elements are _____ to the functional axis.

22. Concentricity is a _____ control that includes eccentricity, circularity, cylindricity, straightness, etc.

23. The tolerance zone for concentricity is always a _____ _____.

24. Circularity tolerance provides a tolerance zone for the feature _____ to lie in.

25. Circularity is a cross-sectional _____ control.

Referring to the drawing in Figure 9-6, answer questions 26–30.

26. Surface profile is a _____ tolerance unless otherwise specified.

27. Bilateral tolerances are assumed to be divided equally on either _____ of the profile unless the designer specified something else.

28. Line profile tolerances are usually specified as a refinement for _____ profile tolerances.

29. What is the total amount of variation permitted with the surface profile tolerance specified? _____

30. Profile tolerances are usually specified for _____ or _____ surfaces.

Hole Pattern—Single Control

Referring to the drawing in Figure 9-7, answer questions 31–38.

31. What is the hole tolerance at LMC? _____ MMC? _____

Figure 9-7

32. How close together can two of the holes get? _____

_____

33. Must the axis of the hole always be within the allowable tolerance zone for that hole? _____

34. Datum "A" controls the _____ or _____ of the holes as they pass through the part.

35. What determines the location of this pattern? _____

_____

36. Must each hole tolerance zone be at the specified basic dimension from each other? _____

37. What is the hole size at MMC? _____ LMC? _____

38. What size would the gage pins be if a gage were made to check these holes? _____

41. Each hole in the pattern _____ shift or move within the pattern that is the specified tolerance.

42. The axis of each feature within the pattern _____ be the basic dimension apart.

43. Each feature within the pattern is permitted _____ tolerance at MMC.

44. The .010 zone _____ be within or at least make line to line contact with the .030 zone.

45. The .010 tolerance is specified to _____ feature-to-feature relationship closer and to control feature _____.

46. The actual feature axis for each hole must simultaneously _____ through both tolerance zones.

47. The feature tolerance zone could range in size from _____ to _____ depending on actual mating size.

Hole Pattern—Composite Control

Referring to the drawing in Figure 9-8, answer questions 39–47.

39. The _____ tolerance is .030 diameter at MMC.

40. Each hole in the pattern is controlled with a _____ diameter tolerance at MMC.

Noncylindrical Features

Referring to the drawing in Figure 9-9, answer questions 48–52.

48. What is the tab MMC? _____

49. What is the shape of the tolerance zone? _____

_____

Figure 9-8

Figure 9-9

50. Does the tolerance control the amount tab angularity in relation to the part axis? _____

51. Why is datum "A" modified in the feature control frame? _____

52. Does the basic dimension, 6 places by 60 degrees, have a tolerance? If so, what is it? _____
_____

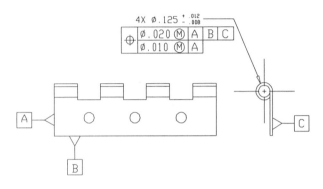

Figure 9-10

Coaxial Features

Referring to the drawing in Figure 9-10, answer questions 53–57.

53. This feature control frame is referred to as a _____ feature control frame.

54. The .020 tolerance provides a _____ tolerance zone.

55. The pattern coaxiality is held to _____ at MMC.

56. The .010 tolerance controls _____ feature in the pattern.

57. Why are the modifiers left off the datum reference letters? _____

APPENDIX A

Past Practices

This Appendix is provided to illustrate past practices. There are many changes and updates to the standard as we know it today; most of these changes are simplifications of various symbols. Those past practices are fairly straightforward and understandable if found on today's drawings. There is one practice, however, that may not be as well understood. This practice is plus/minus positional tolerancing with implied datums.

Plus/Minus Positional Tolerancing

Many existing drawings throughout industry have features and patterns of features located with plus/minus tolerances following dimensions from implied datums. Then features within the pattern were located with basic dimensions and positional tolerances. This combination method will be explained and illustrated below. The part used is the same as that in Figure 8-29 (except dimensioning is changed to the 1973 Standard), where composite positional tolerance was illustrated and explained.

APPLICABILITY OF MODIFIERS

The current standard was revised to where the RFS symbol is no longer required for a tolerance of position. Both of the former Rules Two and Three have been replaced with a new Rule Two. Former Rules Two and Three were:

a) **Rule Two.** *Tolerances of Position. RFS, MMC, or LMC must be specified on the drawing with respect*

Figure A-1

Figure A-2

to the individual tolerance, datum reference, or both, as applicable.

b) **Rule Three.** *All Other Geometric Tolerances. RFS applies with respect to the individual tolerance, datum reference, or both, where no modifying symbol is specified. MMC must be specified on the drawing where it is required.*

TANGENT RADII

The definition of the tolerance zone for the former term *tangent radius*, previously noted by the symbol R, is now meant to apply to a *controlled radius* CR.

DATUM FEATURE SYMBOL

The former datum feature symbol consisted of a rectangular frame containing the datum identifying letter preceded and followed by a dash.

PROJECTED TOLERANCE ZONE

The former method of indicating a projected tolerance zone is illustrated in Figure A-2.

The current standard was revised to where the pro-

jected tolerance requirement is included in the feature control frame (see Figure A-3).

SYMBOL AND DEFINITION

The position symbol and definition are the same for the current practice as they were for the previous practice. Refer to Figure 8-12 for the symbol and the paragraph following the figure for the definition.

APPLICATION

The drawing that will be used to explain and illustrate the combination of positional and plus/minus tolerancing is the same as Figure 8-29 with the dimensioning changed to the 1973 standard, as illustrated in Figure A-4. Note that the plus/minus pattern-locating dimensions are not from specified datums. Dimension origins were implied by the surfaces where the dimensions originated.

PATTERN-LOCATING TOLERANCE

The plus/minus tolerance created a square pattern-locating tolerance zone twice the size of the plus/minus tolerance. The plus/minus tolerance in this example is .010 in. The square tolerance zone would measure .020 in. in the X and Y directions. Refer to Figure A-5 for an illustration of the pattern positional tolerance zone.

FEATURE TOLERANCE

These square tolerance zones provide the zone for each individual feature's circular positional tolerance zone. The center of the positional tolerance zone must

Figure A-3

Figure A-4

.010 DIA. AT MMC POSITIONAL
TOLERANCE ZONE.
(CENTER OF POSITIONAL ZONE
MAY BE ON THE BOUNDARY
OF PATTERN LOCATING
TOLERANCE ZONE.)

Figure A-5

FEATURE LOCATION

The axis of each feature must lie somewhere within the circular tolerance zone. This means that the actual feature axis could lie outside the boundary of the square pattern positional tolerance zone. When the positional tolerance is applied on an MMC basis, any feature not at MMC is permitted an increase in positional tolerance, which could allow the feature axis to be farther from the pattern tolerance zone boundary. The tolerance zone increase, as the feature departs from MMC, is determined the same for the 1973 ANSI standard and ANSI Y14.5M-1982. Figure A-6 illustrates how the features may be positioned within the circular positional tolerance zone.

lie on or within the boundaries of the square pattern plus/minus tolerance zone. The centers of the circular positional tolerance zone must also be at the specified basic dimension from each other. The two tolerance zones are illustrated in Figure A-6.

AXIS OF EACH FEATURE
(TOLERANCE ZONES AT .010 MMC)

Figure A-6

REGARDLESS OF FEATURE SIZE

The ANSI Y14.5M 1982 Standard required the RFS symbol (see Figure A-7) to be specified in the feature control frame when the design required the individual tolerance, datum reference, or both to apply regardless of feature size. Regardless of feature size is implied in the ASME Y14.5-1994 Standard if it is not specified.

Figure A-7

Appendix B

Symbols

Arc Length	1̅0̅5̅
Maximum Material Condition MMC	Ⓜ
Least Material Condition LMC	Ⓛ
Projected Tolerance Zone	Ⓟ
Circularity	○
Straightness	—
Cylindricity	⌭
Perpendicularity	⊥
Flatness	▱
Profile of a Surface	⌓
Profile of a Line	⌒
Counterbore or Spotface	⌴
Depth or Deep	⤓
Dimension Origin	
Quantity	X
Countersink	⌵
Square	□
Reference	(50)
All Around	⌀⟋
Conical Taper	⊳
Flat Taper Slope	⟍
Radius	R
Spherical Radius	SR
Spherical Diameter	SD
Diametrical Symbol	Ø

Basic or Exact Dimension	[2.125]
Datum Feature Symbol	[A]
Feature Control Frame	[⊕ | Ø.010 Ⓜ | A | B]
Datum Target	A1
Parallelism	//
Angularity	∠
Position	⊕
Concentricity	◎
Circular Runout	⟋
Total Runout	⟋⟋
Symmetry	≡

Glossary

Actual local size The value of any individual distance at any cross-section of a feature.

Actual mating envelope This term must be defined according to the type of feature:

a) An external feature—A similar perfect feature counterpart of smallest size that can be circumscribed about the feature so that it just touches the surface at the highest point.

b) An internal feature—A similar perfect feature counterpart of largest size that can be inscribed within the feature so that it just touches the surface at the highest point.

Actual mating size The dimensional value of the actual mating envelope.

Actual size The measured or produced size of a feature.

Allowance The tolerance that permits two parts to be assembled with either a clearance or interference fit between parts.

Angularity The condition of a plane or axis at a specified angle, other than 90°, from a datum plane or axis.

Axis The theoretically centered line through the center of a cylindrical feature.

Barreled The condition of a cylindrical feature where it bulges in the center and tapers down to the ends.

Basic dimension (BSC) A theoretically exact value specified on a drawing to describe the size, shape or location of a feature. Variations from these dimensions are specified as feature tolerances, notes, or other dimensions.

Basic size The size specified for a feature in which the size tolerances apply.

Bilateral tolerance A tolerance specified by the designer that permits a dimension variation in two directions, e.g., plus or minus 0.5.

Boundary, inner A worst-case boundary (locus) generated by the smallest feature (considering the modifiers) minus the stated geometric tolerance and any additional geometric tolerance (if applicable) from a feature's departure from its specified material condition.

Boundary, outer A worst-case boundary (locus) generated by the largest feature (considering the modifiers) plus the geometric tolerance and any additional geometric tolerance (if applicable) from the feature's departure from its specified material condition.

Boundary of perfect form The condition (envelope) of true geometric form represented by the drawing.

Boss A raised area on castings or forgings to allow more material for threading, stop, support area, or bearing area.

Callout A specific note on the blueprint stating dimensions, tolerances, geometric controls, or feature specification.

Center plane The plane or line located in the middle of a noncylindrical feature.

Characteristic An integral part (a symbol) of the geometric dimensioning and tolerancing system or a feature of a part or an assembly.

Checking fixture A go and no-go type gage used to verify features.

Circularity The surface condition of a feature (cone, sphere, cylinder) where all elements in a cross-sectional measurement must be an equal distance from the axis within the limits of a specified tolerance during one complete revolution.

Circular runout The condition of a circular line element during a complete revolution where composite control in relation to the axis is measured at any cross-section along the controlled surface.

Clearance The maximum intentional difference between the size of mating parts.

Coaxiality The condition where two or more axes are in alignment with each other.

Combined feature control frame A feature control frame made up of two or more feature control frames each with a geometric characteristic symbol or a combination of feature control frame and datum feature symbols.

Composite feature control frame A feature control frame made of two or more feature control frames with the same geometric characteristic symbol.

Concentricity The condition where the median points of all diametrically opposed elements of a feature of revolution (or correspondingly located elements of two or more radially disposed features) are in alignment with the axis (or center point) of a datum feature.

Conical Cone-shaped part of features.

Contour tolerancing See Profile of a line or Profile of a surface.

Control A limitation specified by the designer for various features. The limitation is specified as a size or geometric tolerance.

Controlled radius tolerance A tolerance zone defined by two arcs (the minimum and maximum radii) that are tangent to the adjacent surface.

Coordinate A set of numbers used to specify or determine the location from an X and Y axis of a point, line curve, or plane.

Coordinate measuring machine (CMM) An electronically controlled machine used to determine the location or condition of features in space. The data generated may be printed out with a printer option.

Coplanary The condition of two or more surfaces in one plane.

Counterbore A straight-sided recessed area

around the end of a hole so a fastener is flush below the surface of a part.

Countersink A taper-sided recessed area around the end of a hole so a beveled fastener is flush below the surface of a part.

Cylindricity A condition of a surface where all points on that surface are of equal distance from the axis of that surface.

Datum A theoretically exact dimension origin. A datum may be a point, axis, line, or plane used for repeatable measurements.

Datum axis The axis established by the intersection of the X and Y datum planes of cylindrical features, or the axis of a cylindrical feature established by the actual irregularities of the feature extremities. The axis is theoretical.

Datum feature The actual part feature used to establish the datum.

Datum feature symbol A box that contains the datum identifying letter with a dash on either side of the letter.

Datum line The line derived from the actual counterpart of the datum feature. The line may be established by two planes, or the simulated center of cylinders or hole.

Datum of size Any feature specified as a datum reference that is subject to size variation based on plus/minus tolerances such as widths or diameters.

Datum plane A theoretically exact plane established with checking fixtures or gages when in contact with the counterpart of actual datum features.

Datum point A theoretically exact point specified with a datum target of little or no size that has position on a surface for functional gaging purposes.

Datum reference A feature specified on the drawing as a datum feature.

Datum reference frame The three mutually perpendicular planes used to establish the theoretical datum planes for repeatable orientation from design to inspection.

Datum surface A theoretical surface such as cylinders, slots, holes, edges, surfaces, etc. Used to establish repeatability.

Datum target A specified line, point, or area identified with a datum target symbol to establish repeatability.

Diameter Describes through the axis measurement of cylindrical features and tolerance zones.

Dimension The numerical value specified for the size and/or location of features.

Dimension line The line drawn from the dimension value to the feature extension line.

Eccentric A condition where two or more features do not have a common axis.

Error A variation from a desired dimension or geometric form, location, or orientation that is unintentional. Errors are acceptable within the limits of the specified tolerances.

Extension line The line used to extend the object line of parts.

Feature The term given to any physical portion of a part, e.g., surface hole or pin.

Feature, functional A feature controlled geometrically to meet specific design requirements. They are used to locate features from, and/or mate with other features that are interrelated in the overall design.

Feature control frame A rectangular box containing the specific instruction(s) for a feature or group of features. The rectangle contains the geometric characteristic symbols, tolerance, datum reference letters, and maybe the diameter and modified symbols.

Feature of size One cylindrical or spherical surface, or a set of two opposed elements or opposed parallel surfaces, associated with a size dimension.

Feature-relating tolerance zone framework The upper half of a composite feature control frame that provides the location of feature patterns as well as interrelation of features within these patterns.

Fit The clearance or interference between two mating parts or fasteners.

Flatness The condition of a surface having all elements in one plane.

Form The desired finished shape for a given feature.

Form tolerance A tolerance specified to allow a specified variation in a feature or surface from the desired perfect form.

Free-state variation Describes the distortion of a part after removal of forces applied during manufacturing. Parts are referred to as a nonrigid part.

Full indicator movement (FIM) This term replaces older terms, Full Indicator Reading (FIR) and Total Indicator Reading (TIR). It is the full movement of an indicator needle while measuring a feature during a full rotation or complete travel along a feature.

Function The consideration for the movement involved with assembled parts.

Geometric characteristics The symbols specified for form orientation, profile, runout, and location tolerancing. These form the symbolic language of GD&T.

Implied datum An unspecified feature implied by dimensioning to be the origin for measurements.

Indicator A precision measuring tool used when checking feature variations.

Knee A piece of equipment used to rest parts against during machining and inspection operations.

Leader The line or arrow used to tie dimensions, symbols, etc., to part features.

Least material condition (LMC) This term is a modifier that is specified with tolerances and datums. It means the condition in which a part or feature contains the least amount of material; for example, the smallest pin size or the largest hole size.

Limit dimensioning The specified maximum and minimum sizes of a feature.

Limits of size The extreme minimum and maximum sizes permissible for a feature when considering the tolerances.

Location tolerance A tolerance specified to allow a specified variation in the perfect location of a feature(s) as drawn on the drawing. The tolerance applies in relation to the datum features used to locate the controlled feature(s).

Maximum dimension The acceptable upper limit or the largest value specified for a feature.

Maximum material condition (MMC) This term is a modifier that is specified with tolerances and datums. It means the condition in which a part or feature contains the maximum amount of material; for example, the largest pin size or the smallest hole size.

Median plane The center or middle plane of a noncylindrical feature.

Minimum dimension The acceptable lower limit or the smallest value specified for a feature.

Minimum material condition See Least material condition.

Modifier In GD&T, the term is used to modify or change specified feature tolerances. There are three modifiers: LMC, MMC, and RFS.

Multiple datum reference frames Referred to when more than one datum reference is specified on the same part.

Nominal size The stated designation that is used for the purpose of general identification of material, or the theoretical size of a feature, e.g., 1-in. bar or 1.000 plus or minus .010.

Optical comparator An instrument for comparing a surface with an ideal surface or standard.

Orientation The relationship of one feature as it relates to another specified as the datum.

Orientation tolerance This tolerance is applicable to features that are related to another feature. The orientation controls are perpendicularity, angularity, and parallelism. Each of these must be related to a datum feature.

Origin The location where dimensions and tolerances begin.

Parallelism The condition of a surface or axis at an equal distance from all points of a datum surface or axis.

Pattern A particular arrangement of features in a part.

Pattern-relating tolerance zone framework The lower half of a composite feature control frame that provides the smaller positional tolerance for each feature within the pattern.

Perpendicularity A condition of a feature median plane, axis, or surface that is 90° from a related datum axis or plane.

Pitch The distance from a thread point to the corresponding point of the next thread.

Plane A surface condition that is straight, flat, or level.

Position A location described by dimensions or the actual location of a feature.

Positional tolerance See Location tolerance.

Primary datum A theoretical plane requiring three points of contact to establish that plane on an actual part. First in importance.

Profile of a line The condition where a permitted amount of profile variation, bilaterally or unilaterally along a line element, is specified.

Profile of a surface The condition where a tolerance zone is established by a three dimensional zone extending the length and width, or circumference of the considered feature.

Projected tolerance zone A zone specified for a given height above a hole in which a pin, cap screw, bolt, etc., is to be installed. The zone provides a cylindrical area in which the feature axis must lie. It controls the perpendicularity of the hole to prevent interference with the mating part.

Radius Describes a measurement from the axis to the surface of cylindrical features.

Reference dimension A dimension specified as an approximate dimension that is not tolerances.

Regardless of feature size (RFS) This term is a modifier that is specified with tolerances and datums. It means that, regardless of the actual feature or datum of size, the stated feature tolerance applies.

Relationship The consideration of an assembly in a static or fixed situation.

Roundness See Circularity.

Runout The composite variation from a desired surface during a complete rotation of the part around the axis.

Runout tolerance A specified variation for an actual feature surface or line in relation to the axis during one complete revolution.

Secondary datum A theoretical plane requiring two points of contact to establish that plane on an actual part. Second in importance.

Size, feature of See Feature of size

Size, resultant condition The actual value of the resultant condition boundary.

Size tolerance A tolerance specified to allow a feature to vary a specified amount.

Slope The inclination of a surface expressed as a ratio of the difference in heights at each end divided by the distance between those heights.

Specification A detailed description of a feature requirement such as size, shape, location, tolerance, etc.

Specified datum A datum feature identified with a datum feature symbol.

Spherical A globular body, ball, having all points equal distance from a given center point.

Spot face A machined surface of a specified size on a casting or forging.

Squareness See Perpendicularity.

Statistical tolerancing Assignment of tolerances to related components of an assembly on the basis of sound statistics.

Straightness The condition of a surface or axis where a single line element must be a straight line.

Symmetry The condition where the median points of all opposed or correspondingly located elements of two or more feature surfaces are in alignment with the axis or center plane of a datum feature.

Tertiary datum A theoretical plane requiring one point of contact to establish that plane on an actual part. Third in importance.

Three-plane concept The concept of three mutual planes exactly theoretically 90° and perpendicular to each other. Used for repeatable orientation.

Tolerance The total amount a specific dimension may vary. Between the maximum and minimum limits of size.

Tolerance zone A boundary established by a size or geometric tolerance in which the actual feature must be contained.

Total runout The condition of all surface elements during a complete revolution where composite control in relation to the axis is measured.

True position An exact (perfect) location described by basic dimensions in relationship to a datum or other feature. The location specified on the drawing.

Unilateral tolerance A tolerance that permits variation in one direction from the specified dimension, for example, 1.000 plus .000, minus .010.

Virtual condition The condition of a feature where the collective effect of size and form error establishes the feature size required for determining fit between parts.

Weldment A unit or assembly formed by the welding together of pieces.

Wristed The condition of a cylindrical feature where it narrows toward the center from each end.

ANSWERS TO QUESTIONS

Chapter One

Answers: **1.** communication; **2.** symbols; **3.** American Society of Mechanical Engineers (ASME); **4.** clarity; **5.** replace; **6.** total; **7.** form; **8.** size, location; **9.** function, relationship; **10.** tolerances, interchangeability; **11.** tolerance; **12.** plus/minus

Chapter Two

Answers: **1.** 17; **2.** 2; **3.** 5; **4.** 10; **5.** 19; **6.** 6; **7.** 5; **8.** 1; **9.** 22; **10.** 20; **11.** 7; **12.** 3 & 4; **13.** 8; **14.** 21; **15.** 11; **16.** 14; **17.** 23; **18.** 12; **19.** 15; **20.** 16; **21.** 9; **22.** 6; **23.** 17

Chapter Three

Answers: **1.** specified; **2.** dimension, orientation; **3.** surface (planes), axis, center lines, edges (lines), points, areas; **4.** no; **5.** three; **6.** 90; **7.** circle; **8.** size; **9.** lines, points, areas; **10.** tolerance; **11.** simulated; **12.** modifiers; **13.** convey; **14.** rock; **15.** flat

Chapter Four

Answers: **1.** control; **2.** combined; **3.** surface; **4.** geometric characteristic; **5.** tolerance; **6.** left, right; **7.** primary; **8.** line; **9.** compartments; **10.** no

Chapter Five

Answers: **1.** size, form; **2.** individual; **3.** MMC or LMC; **4.** MMC; **5.** zero; **6.** RFS; **7.** pitch; **8.** beneath; **9.** foundation; **10.** not

Chapter Six

Answers: **1.** directions; **2.** radially; **3.** element; **4.** boundaries; **5.** entire; **6.** error; **7.** tolerance; **8.** diameter; **9.** attachment; **10.** implied; **11.** assumed; **12.** size, geometric; **13.** axis; **14.** distance; **15.** width; **16.** RFS; **17.** bilateral; **18.** line; **19.** profile; **20.** revolution

Chapter Seven

Answers: **1.** collective; **2.** geometric; **3.** exceed; **4.** gages; **5.** acceptable; **6.** mating; **7.** opposite; **8.** interchangeable; **9.** Four; **10.** modifier

Chapter Eight

Answers: **1.** position, concentricity; **2.** variation; **3.** RFS; **4.** cylindrical, noncylindrical; **5.** basic; **6.** specified; **7.** increases; **8.** shift/rotation; **9.** axis; **10.** length, depth; **11.** virtual condition; **12.** MMC; **13.** MMC; **14.** size; **15.** noncumulative

Chapter Nine

Answers. **1.** D = .005, E = .040; **2.** D = .003, E = .003; **3.** RFS; **4.** Yes, Rule 3; **5.** 2.025 × 1.025 × 4.00; **6.** .005; **7.** .002; **8.** Yes, Rule 1; **9.** .475 × 1.9751 **10.** No; **11.** datum feature; **12.** noncylindrical—the diameter symbol is not specified preceding the .015 tolerance; **13.** width; **14.** Yes, parallelism; **15.** surface; **16.** total; **17.** FIM; **18.** total; **19.** One; **20.** axis; **21.** irrelevant; **22.** composite; **23.** diameter; **24.** surface; **25.** line; **26.** bilateral; **27.** side; **28.** surface; **29.** 010; **30.** irregular, coplanar; **31.** .040, .030; **32.** 1.462; **33.** yes; **34.** attitude, perpendicularity; **35.** The basic dimensions from datum surfaces "B" and "C"; **36.** yes; **37.** .528 diameter, .538 diameter; **38.** .498 diameter; **39.** pattern; **40.** .010; **41.** may; **42.** must; **43.** .010; **44.** must; **45.** control, attitude; **46.** pass; **47.** .010; .020; **48.** 313; **49.** It is a width; **50.** yes; **51.** Because it is a feature of size; **52.** No. Basic dimensions do not have tolerances. The spacing is permitted to vary based on each feature's tolerance; **53.** composite; **54.** pattern; **55.** .020; **56.** each; **57.** Because they are not features of size, they are all surfaces.

Index